THE
ENGLISH
ROSES

DAVID
AUSTIN

THE
ENGLISH
ROSES

Classic favorites & new selections

SPECIAL PHOTOGRAPHY BY
HOWARD RICE
AND ANDREW LAWSON

FIREFLY BOOKS

CONTENTS

PART

ONE

THE ORIGINS AND

NATURE OF

AN ENGLISH ROSE

PREFACE

ENGLISH ROSES have developed considerably since I wrote my previous book in 1993. Many new varieties have been introduced while others—now considered to be inferior—have been dropped. As a result, the beauty of English Roses, both in flower and growth, has been greatly enhanced. Health and vigour has also been improved.

Not only have English Roses developed in beauty; they have also broadened widely in the range of their beauty, so that as we walk around a garden of English Roses we can always come across something different. As a result of all this, it has become necessary to divide them into six sub-groups, each with its own particular character, so that gardeners may better understand them and enjoy them more completely. These six groups are described for the first time in this book. The new Climbing English Roses are among the most interesting.

With all these developments, it is only natural that we have had new insights into the way in which English Roses can be used in the garden. The greater range opens up further possibilities, all of which are all discussed in Part III.

Creating a book of this kind is inevitably a team effort and I would like to take this opportunity of thanking all those who have made a contribution: Andrew Lawson, Howard Rice, Ron Dakin and others, for their magnificent photographs; Erica Hunningher, not only for editing the book, but for providing so much wise advice; Ken Wilson for his elegant design; and not least, Diane Ratcliff for her work, patience and skill in typing the manuscript and helping me generally in my task.

Finally, I would like to thank my son David J. Austin, without whom English Roses, as we know them today, would not have been possible.

I dedicate this book to my grandchildren
Ellen, Robert, Richard, James, Olivia, Katherine,
Frederick and Charlotte.

PART
ONE

—

THE
ORIGINS
AND
NATURE
OF AN
ENGLISH
ROSE

1

THE
ROSE

THE ROSE, at its best, is the most beautiful of all garden flowers. Polls have been taken and surveys made and these have shown consistently that the rose is the most popular of flowers by a very wide margin. This is in spite of the fact that some of us believe that the roses of the twentieth and twenty-first centuries have lost something of the beauty of their forebears—those known today as the Old Roses.

Why the rose has been so popular over the centuries is not difficult to see, although it is rather more difficult to explain. It is certainly very different from most other garden flowers: a walk around the garden in summer will make this immediately clear. There is a human quality about roses that gives them a certain appeal. We find the same quality in other members of the great *Rosaceae* family—the ornamental cherries and other *Prunus* and so on. It is interesting to note that the rose is close to the peony in appearance and that the peony, not the rose, has a special place in the life of China and Japan. This must say something about the rosette form of flower. Both plants, in their garden form, have double flowers of very similar character. Both flowers are unusually rich in fragrance. However, for all its beauty, the peony is a much more limited flower. It blooms but once in a season and has nothing like the range of growth and form found in the rose.

The rose as a garden plant goes back to at least three hundred years before the birth of Christ and, almost certainly, a great deal further. Theophrastus (c. 370–286BC) wrote, in his *Enquiry into Plants*, of roses having anything from five to one hundred petals. These must have been garden roses, as no wild rose has more than five petals. At about the same time, coins of the island of Rhodes depicted the flowers of a rose. Ever since then the rose has been intertwined in Western culture. Throughout Greek, Roman and Persian history it makes many appearances, sometimes because of its medicinal qualities, sometimes for its beauty.

INSET Roses depicted on a north African tombstone from the 5th century AD.

OPPOSITE Old Roses with other flowers in Jan van Kessel the Elder's *Still Life of Flowers in a Vase*, 1661.

The Origins and Nature of an English Rose

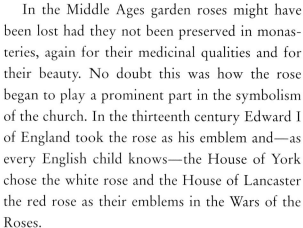

In the Middle Ages garden roses might have been lost had they not been preserved in monasteries, again for their medicinal qualities and for their beauty. No doubt this was how the rose began to play a prominent part in the symbolism of the church. In the thirteenth century Edward I of England took the rose as his emblem and—as every English child knows—the House of York chose the white rose and the House of Lancaster the red rose as their emblems in the Wars of the Roses.

By the eighteenth century varieties of roses were numbered in thousands. These were mainly of French origin, but some were Dutch and others British. In the first decade of the nineteenth century a vast collection of roses was assembled at the Empress Josephine's Malmaison in Hauts-de-Seine, France, and it is she who was said to have set the rose on course for its present prominent position in the garden. By the twentieth century the rose no longer belonged to certain areas and countries, but to the world.

The rose has many virtues, among them a softness that gives it a unique beauty, but I think it is the arrangement and texture of the petals that is the most special quality. The way the light plays through and between the petals creates a vast range of pleasing and ever-changing effects as the flower develops. Not only this, but if we look at roses as a whole we see how the form and character of their flowers differ endlessly from one variety to another. The bloom of a rose has a range of beauty as it develops from the bud to the fall of the petals. It is a very different flower as the day unfolds and from one day to the next, the nature of the day itself, light levels in particular, also contributing to the flower's appearance. It is this multitude of effects—which we do not find to the same degree in any other flower—that helps to make the rose so very special. The peony and, to a lesser extent, the camellia are the only flowers to rival the rose in this respect. In the flowers of a rose we have not one flower, but many different flowers.

The beauty of the rose is not confined to what we see: it is also one of the most fragrant of flowers. Its fragrance lifts our spirits as few others can. Nor is there only one fragrance among roses: it varies greatly between one variety and another and between species. Indeed, the scents of many other flowers, such as lilac and lilly of the valley, are to be found among

The Origins and Nature of an English Rose

roses—even if the so-called Old Rose fragrance is the one we love best.

Not only is the rose the most popular of garden plants, it is also one of the favourite flowers for cutting and arranging in bowls for the house. It is not the most long-lasting when picked, but what other flower can provide so much pleasure while it lasts? In the house we can observe and appreciate it at close quarters. A bowl of roses will light up a whole room.

The humanity of the rose has made it one of the great symbols of Western civilisation and, indeed, of the civilisations of the Middle East. It has long been used as a symbol of feminine beauty and romance—and this tells us something not only about the beauty of women, but about the beauty of the rose too. It is also often used as a symbol of early childhood. Down the centuries it occurs in the visual arts, in literature and in song. It has been possibly the most popular subject for use in decoration—on fabrics, pottery, architecture and so on. It has been used as a symbol in Christianity as, for example, in the rosary. It is the national flower of both England and the United States of America. It appears in British law in the term 'sub-rosa' (literally 'under the rose', meaning privately). It has also been adopted as the symbol of the Labour Party in Britain—and the list continues. The rose is, in fact, more than just a flower; it is part of civilisation.

The uses of the rose are certainly more numerous than those of any other flower. We may grow it as a shrub in the mixed border or in a border only of roses. With its usually sympathetic colours, gentle form and somewhat random growth, it will mingle with other plants as well as with its own kind, both easily and naturally. Roses make colourful hedges and are good bedding plants, if we are careful to select low-growing varieties. They may also be grown as standards to give height to low-growing, formal displays. Then there are the climbers. It is sometimes a little hard to believe that the rose—which excels in so many roles—is also the best of all climbing plants. Nor should we forget species roses, most of which are worthy of a place in the garden. They are ideal for wild areas and have brightly coloured autumn hips.

It is small wonder that the rose, over the centuries, has been proclaimed the 'Queen of Flowers'. Those of us who breed roses have a special responsibility to retain its beauty above all others. I can only hope that my readers find the English Roses worthy of their place in the story of the rose.

OPPOSITE ABOVE Trellis covered with climbing roses, a detail from *Emelye in her garden*, a 15th-century illustration in a French translation of Boccaccio's *La Teseida*.

OPPOSITE BELOW A Mughal courtier holding a rose, a detail from a 17th-century Indian miniature.

BELOW *Rosa Gallica Regalis*, a variety now lost to cultivation, by Pierre Joseph Redouté (1759–1840).

2

THE
IDEA OF
THE
ENGLISH
ROSE

INSET Graham Stuart Thomas (1909–
2003), who did so much to make
possible the reintroduction of Old
Roses. Here, he holds blooms of
the English Rose 'Graham Thomas',
which we named in his honour.

OPPOSITE 'The Generous Gardener'
perfectly illustrates the Old Rose
charm of an English Rose.

ENGLISH ROSES are just one more step in the long journey of the rose. They differ from other contemporary roses in two fundamental ways—the shape of the flower and the growth habit—and combine two great traditions: the full rosette or cup shape, fragrance and general character of the Old Roses and the wide range of colour and repeat-flowering of present-day Hybrid Teas and Floribundas.

For many centuries garden roses had double-petalled flowers arranged in the shape of a cup or rosette, or single or semi-double flowers exposing the elegance of their stamens. Then, in the latter part of the nineteenth century, breeders began to produce roses for their buds alone. These were originally developed in the Tea Roses and the Hybrid Perpetuals and later, more completely, in the Hybrid Teas, which were bred to be short and upright. Hybrid Teas became so popular that by the beginning of the twentieth century the Old Roses—as they eventually became known—had all but disappeared from the garden and the Hybrid Teas reigned supreme.

With the triumph of the Hybrid Teas, sophisticated gardeners began to regret the decline of the Old Roses and they started to collect such varieties as had survived. Among them were Lawrence Johnson of Hidcote Manor Garden in Gloucestershire, Victoria Sackville-West, poet and garden writer, of Sissinghurst Castle in Kent, and Hilda Murrell, who had a nursery near Shrewsbury. After the Second World War their collections were gathered together and added to by the late Graham Stuart Thomas, first at Hillings & Co of Woking, and later at Sunningdale Nurseries, Surrey, which formed the basis

of the National Collection at Mottisfont Abbey in Hampshire. The large number of varieties he was able to offer gardeners rendered a very great service to all who loved roses and indeed, to the gardening public as a whole. In 1956 Graham Thomas became Gardens Adviser to The National Trust, which gave him ample opportunity to bring Old Roses to the fore. Today Peter Beales and his family hold a large collection of Old Roses at their

The Origins and Nature of an English Rose

nursery at Attleborough in Norfolk. We ourselves at David Austin Roses have an extensive collection near Wolverhampton. In addition there are numerous private collectors.

To Old Rose enthusiasts the popular Hybrid Teas and Floribundas had become too 'stumpy' and angular in growth and their flowers too harsh in colour and character. Not least, the rich fragrance of many Old Roses was usually lacking among the Hybrid Teas and Floribundas. However, while Old Roses are both more beautiful and better garden plants, the Hybrid Teas—and more recently the Floribundas—have certain advantages. Foremost among these is the fact that they have been bred to produce flowers repeatedly throughout the summer and not just once in early summer, as is the case with many of the Old Roses. Although the later Old Roses do have this ability, they are never reliably repeat-flowering, nor were they ever developed to anything like perfection. The Modern Roses also have a much wider range of colour than is found in the Old Roses. The colours of the Old Roses are largely confined to white, through soft pink to deep pink, purple and mauve. It is common to think of Old Roses as having crimson flowers, but this colour did not appear in its purest form until the nineteenth century, among the Hybrid Perpetuals. Strong yellow became available only when the French rose breeder Joseph Pernet-Ducher hybridised the Austrian briar with the Hybrid Teas some time around the year 1900. In so doing he bestowed a great gift upon the rose for which he deserves our gratitude. It is true that the old Tea Roses did previously include some

OPPOSITE 'A Shropshire Lad' is an English Rose with flowers beautifully held on shrubby growth.

BELOW LEFT 'Belle de Crécy,' one of the most beautiful of the old Gallica Roses, a group which contributed so much to the breeding of the English Roses.

BELOW RIGHT The charming 'Maiden's Blush' (prior to 15th century) is one of the old Alba Roses which were used in the breeding of English Roses.

varieties in shades of yellow, but these tended to be pale and few in number.

I developed the English Roses with the best qualities of the old and the modern in mind. If the fineness of the old and the youthfulness of the modern were combined in one class of rose by means of hybridisation, I felt that it might be possible to create a group of roses that would be superior to both. This was not as easy as it might sound. It was difficult to get the required balance: when we achieved the colour and repeat-flowering we wanted from our Modern Roses, we would often lose some of the essential beauty

of the flower and the growth of the Old Roses—and so it was with many other factors. But eventually, by careful selection from many thousands of seedlings, the kind of rose we required began to emerge.

One problem was the fact that many of our early roses, though beautiful, were weak and somewhat subject to disease. This has since been overcome and our more recent varieties are more robust and disease-free than the majority of Hybrid Teas and Floribundas—and indeed than most other earlier

groups of roses. That we have been able to achieve this is in some degree due to the fact that we had a far wider range of original parents to work with than was the case with the Hybrid Teas and Floribundas. The Old Rose style of flower that we chose for English Roses was closer to the natural character of wild roses. It was therefore possible to draw upon a greater number of different garden and species roses, each of which had the potential to bring in desirable characteristics. Our overall aim has been, quite simply, to produce a more beautiful rose on a more beautiful shrub. At the same time we have endeavoured to create a more healthy and robust plant.

Today some people in certain countries refer to our roses as 'David Austin Roses'. We prefer to call them 'English Roses', not for any nationalistic reason, but because it seems to us that England, more than any country, is associated with gardens—and, more particularly, with the rose itself; although the French might argue with this. Just as we have Japanese chrysanthemums and peonies, French marigolds and Dutch and German irises, so it seemed reasonable to call our roses 'English'. It has been my hope that breeders from all countries will eventually take up this group and help in their development—and, thus, a new class of roses will be born and their future ensured.

ABOVE LEFT Warm Wishes ('Fryxotic'), a Hybrid Tea.
ABOVE RIGHT Margaret Merril ('Harkuly'), a Floribunda.
Roses of these two classes provided the repeat-flowering character for the English Rose.

OPPOSITE 'The Countryman' is one of our earlier English Roses of typically Old Rose character.

The Origins and Nature of an English Rose

THE ANCESTORS OF THE ENGLISH ROSE

No plant breeder could have been blessed with so wide a range of parent material as I was when I first set out to breed English Roses in the early 1950s. No other major garden flower that I am aware of—at least in the West—has so long a history as the rose. Not many flowers have so many wild species of true garden value. Few flowers that I can think of have so wide a range of garden classes and groups, each with their own particular beauty. Not many other flowers have so many garden varieties.

Various groups of roses, dating from the earliest times and to the present day, have played a part in the breeding of English Roses. During the development of the English Roses we were frequently held up by the fact that certain roses made poor parents: they either failed to produce seed or the seed failed to germinate—and when they did, they often failed to hand on their good qualities to their offspring. Only by making many crosses and rearing hundreds of thousands of seedlings and continually selecting from these have we succeeded in producing the English Roses.

I start with the so-called 'Old Roses', none of which can be said to be the result of any kind of planned hybridisation—with the possible exception of the Hybrid Perpetuals. Their development was a matter of sowing a quantity of seed and seeing what turned up. If a certain type of rose appeared that was sufficiently different or superior in any way, it was preserved and introduced into the garden.

The Old Roses

First and foremost among the original parents of the English Roses are those groups that I have chosen to call the *true* Old Roses; that is to say, those that are summer-flowering only and existed long before the arrival of the repeat-flowering roses of the nineteenth century. None of these groups can be said to be completely distinct—the breeding of each has become in some degree entangled with the others. Nonetheless, each has its own special overall character.

There are five groups of Old Roses: the Gallicas, the Damasks, the Albas, the Centifolias and the Moss Roses. Representatives from the first three of these groups have been used in the breeding of English Roses.

The Gallica Roses are perhaps the most important group among the Old

OPPOSITE *Rosa × alba* 'Alba Maxima' is an Old Rose which forms a fine large shrub. Like all the original garden roses, it flowers only once in a season.

The Origins and Nature of an English Rose

Roses and it is here that we find most of the mauves, purples and purple/crimson shades. These colours are often particularly rich and pleasing. The Gallicas have rather short growth by Shrub Rose standards. Despite their ancient origin, they are excellent garden plants to this day. Unfortunately, in comparison with other Old Roses, they are not always very fragrant—although some of them are. They do, however, have the true Old Rose beauty in a high degree. They often have beautifully formed flowers. They are extremely tough and hardy and it is probably for this reason that so many of them have survived for so long. All these qualities they have brought to the English Roses.

The Damask Roses are a group of great antiquity. They originated in the Middle East and were said to have been first brought to Europe by the Crusaders. They usually have lovely, airy, rather open growth and elegant, light green foliage with widely spaced leaflets. The flowers are generally of lovely, glistening, clear pink colouring—although there are one or two mauves and purples, as well as a white variety, 'Madame Hardy'; all of which are partly of hybrid origin. They have the true, rich Old Rose fragrance—which is, in fact, sometimes known as the Damask fragrance. They have passed on this fragrance, together with something of their elegance, to many of the English Roses—sometimes via the Portland Roses, of which they are a parent.

The Alba Roses are, to my mind, among the most beautiful of the Old Roses. They are believed to have been the result of chance crosses between the Damask Rose and *Rosa canina*, the dog rose. Albas are larger in growth than most Old Roses—up to 180cm/6ft or more in height. They are less inclined to shoot from the base of the plant and tend to build up growth on existing branches. For this reason they were at one time known as 'tree roses', although this seems hardly accurate. Their flowers often have a pleasing formality and they have a delicate charm which is very much their own. Their colour range is limited to white, through blush pink, to pink. They also have a lovely fragrance. Their foliage is similar to that of a dog rose but of an attractive greyish-green colour. Overall the Albas are particularly strong and healthy and have been responsible for a whole sub-group of English Roses, which we call the English Albas.

The Centifolia Roses, sometimes known as Provence Roses, appeared on the scene much later than the other three groups. Their ancestry is the result of the mingling of various roses. They usually have large, voluptuous flowers which may be of cupped or rosette shape—and they nearly always have a rich fragrance. We have used the Centifolias only occasionally as parents in the English Roses, and then with little success, and we have not used the closely related Moss Roses at all.

These, then, are the five groups of Old Roses that have given people so

The Origins and Nature of an English Rose

much pleasure over the ages and they are still excellent shrubs by present-day standards. The roses that followed them were more in the nature of a lead-up to the 'Modern Roses'.

The Repeat-flowering Old Roses

At the end of the eighteenth century and in the early part of the nineteenth century, four roses were brought to Europe from China. These were destined to have a profound effect on the future of the rose, not least because they flowered not just once in early summer, but repeatedly throughout the summer. They have been called the Stud Chinas. There were four: 'Slater's Crimson China' (1792), 'Parsons's Pink China' (1793), 'Hume's Blush Tea-scented China' (1809) and 'Parks's Yellow-scented China' (1824). These slowly became interbred with European roses, so that by the end of the nineteenth century nearly all roses that were introduced had the ability to repeat-flower.

'Slater's Crimson China' was of particular interest, since it bore flowers of a rich, unfading crimson—a colour that had not been present in

European roses up to that time. This colour was eventually passed on—via a somewhat tangled path—to the roses of the present day. However, the effect of the China Roses upon garden roses was much more significant than their colour—they changed the character of roses. Being generally less heavy and more twiggy in growth and having polished leaves of a darker green, their whole appearance was more polished. It is only necessary to compare an Old Rose with a Modern Rose to see how great a change this was. So it was that we eventually came to have two entirely different groups of roses—the Old and the Modern.

The first of the new repeat-flowering roses were the Portland Roses, which were the result of crosses between the Damask Rose 'Quatre Saisons' and the China Rose 'Slater's Crimson China'. 'Quatre Saisons' was a repeat-flowering Damask—in fact, the only repeat-flowering rose up to the time of the China Roses—and so the Portland Rose has repeat-flowering characteristics on both sides of its parentage. Unlike the groups that followed them, the Portland Roses were, on the whole, closer to the Old Roses than the Chinas—in both flower and leaf. Not many varieties have survived, and probably not many were bred. Nonetheless, they include varieties with beautiful, rosette-shaped flowers, often in rich, glowing shades of pink, and usually also have a particularly rich Old Rose fragrance. These qualities, combined with compact growth and some ability to resist disease, make them excellent plants that are still well worth a place in the garden. Their contribution to English Roses has been considerable.

The Portland Roses' period of popularity was cut short by the arrival of more dramatic introductions on the garden scene. The first group to appear was the Bourbons. These were largely the result of crossing China Roses with various Old Roses, usually via a minor group known as the Hybrid Chinas. Their foliage and general appearance has much in common with modern varieties, but the flowers retain the Old Rose shape, often deeply cupped, and nearly all are very fragrant. They are also shrubby in growth. The Bourbons, in fact, might be said to have much in common with the English Roses. No sooner had they gained popularity, however, than they were overtaken by the Hybrid Perpetuals.

The Hybrid Perpetuals provide the link between the Old Roses and the Modern. We can see in them the beginnings of the Hybrid Tea Rose, which expresses its beauty in the bud at the expense of the mature flower. Having

TOP Portland Rose 'Jacques Cartier' (prior to 1809) is one of the early repeat-flowering Old Roses.

ABOVE 'Madame Isaac Pereire' (1881) is one of the old Bourbon Roses which, together with the Hybrid Perpetuals, forms a link between the Old Roses and the Modern Hybrid Teas.

The Origins and Nature of an English Rose

large, heavy flowers, the Hybrid Perpetuals were taken up by gardeners whose hobby was to produce blooms for the show bench, and these are not necessarily the best roses for the garden. They include some beautiful varieties but as a group represent a decline in the beauty of the rose. The flowers can be a little coarse and the plants themselves tend to be ungainly. These roses, however, had two great virtues: they were often richly fragrant and they were the first class of roses to include any quantity of varieties with flowers of pure rich crimson colouring. (The Gallica Roses usually have dark flowers, but the colours are closer to mauve or purple, the sole exception being 'Tuscany' whose colour comes very close to a pure crimson.) The Hybrid Perpetuals have played no more than a very small part in the development of English Roses.

While the Hybrid Perpetuals were in vogue in late Victorian times, another group of roses appeared alongside them. These were the Tea Roses, which hold a position somewhere between the Old and the Modern Roses. They were a development of the China Roses, some of which have in their

TOP 'Baroness Rothschild' (1868), now 'Baronne Adolph de Rothschild', is one of the Hybrid Perpetuals that led to the Modern Hybrid Teas. It shows the first signs of a short rose suitable for growing in beds.

ABOVE 'Lady Hillingdon' (1910) is one of the Tea Roses which, when hybridised with the Hybrid Perpetuals, brought the long pointed bud to the Hybrid Teas.

LEFT The species rose, *R. gigantea*, is one of the ancestors of the Tea Roses.

BELOW LEFT Savoy Hotel
  ('Harvintage'), is one of the Hybrid
  Teas that were the dominant roses
  of the twentieth century.
BELOW RIGHT 'Miss Edith Cavell'
  (1917), one of the Dwarf Poly-
  anthas which were parents of
  the Floribunda Roses.

make-up an element of *Rosa gigantea*, which is a very large-flowered species with long, pointed buds. These qualities were passed on to the Tea Roses, which have a delicate beauty, often with the Hybrid Tea type of bud and soft colours and slender, twiggy growth. They also have what is known as a 'tea' scent, similar to the aroma from a newly opened packet of China tea. They are subject to frost damage. We have not used these roses in our breeding, although their influence comes through via other groups.

The Modern Roses

The so-called 'Modern Roses' complete the equation in the breeding of the English Roses. Most important are the Hybrid Teas, which were the result of hybridising the Hybrid Perpetuals with the Tea Roses in the 1840s. The Hybrid Teas are an entirely new type of rose bearing so little resemblance to those of the past that they might almost be regarded as a new plant. For the best part of the twentieth century they dominated the rose scene and are still the most popular roses today. In fact, in Britain, there are few gardens which do not contain at least one or two examples of the Hybrid Tea Rose. Bred primarily as bedding roses, as the years went by, they began to be

grown largely in rose beds. The Hybrid Tea, although beautiful in the bloom, has never been a very good border plant. Nonetheless, it has made a useful contribution to the English Roses, largely because of its wide range of colours and its ability to repeat-flower well, although the shape of its flowers has often proved a problem—the bud shape having a tendency to turn up among our seedlings, which of course was not what we wanted.

The next important group to arrive on the garden scene were the Floribundas in the early years of the twentieth century, for which we have to thank Poulsen of Denmark. These were hybrids between the Hybrid Teas and the Dwarf Polyantha Roses. The latter, which produce small pompon flowers in large sprays, are a small group of very hardy, bushy roses that are, in effect, a dwarf version of the Multiflora Ramblers. The Floribundas have a great deal in common with the Hybrid Teas, except for the fact that in the main they disposed of the Hybrid Tea bud and are largely notable for their massed colour effect. They are more floriferous, more vigorous and more hardy than the Hybrid Teas. The Floribundas have played a more important part in the development of English Roses than the Hybrid Teas, largely because they have more open flowers—and are thus nearer to an Old Rose in appearance—and because they are more free-flowering, vigorous, hardy and disease-resistant. It is unfortunate that the Hybrid Teas and the Floribundas have begun to merge, so that it is sometimes hard to differentiate between the two.

For most of the time that the Hybrid Teas have been in vogue, another group of roses has been slowly evolving. These, for the want of a better name, are usually called Modern Shrub Roses, no doubt to distinguish them from the Old Roses which are, of course, also shrubs. They are a very mixed bunch. Some are very good, strong-growing shrubs but others are on the coarse side. They have played a small part in the development of English Roses.

ABOVE Hybrid Teas and Floribundas were bred largely as bedding roses. Freedom ('Dicjem'), a Hybrid Tea, is seen in full flower in Regent's Park, London.

LEFT Valentine Heart ('Dicogle') is an example of the Floribundas that were important parents of the English Roses.

BELOW 'Penelope' (1924) is a Hybrid
Musk Rose, a group that is the result of
crossing Hybrid Tea Roses with roses
of Musk Rose descent.

BOTTOM 'Desprez à Fleurs Jaunes' (1826)
is a member of the Noisette group, the
result of hybridising the Tea Roses with
the Musk Rose to produce a repeat-
flowering climber.

TOP RIGHT 'Hansa' (1905) is a
vigorous Rugosa Rose, a group
bred from *Rosa rugosa*, a naturally
repeat-flowering species, and
notable for extreme hardiness and
disease-resistance.

Alongside the Modern Shrubs we have hybrids of the Rugosa Rose, which are usually placed in a class of their own. *Rosa rugosa* is a strong-growing species that is naturally repeat-flowering in the wild, a quality found in only two other species. The Rugosa Roses are exceptionally healthy and make fine shrubs; though they can be a little coarse in appearance. They are at the present time playing an important part in the development of English Roses. There are also the Hybrid Musks, which are a small group resulting from crosses between the Musk Rose and certain Hybrid Tea and Tea Roses. They are all good shrubs and well worth a place in the garden.

Climbing Roses

All the roses mentioned so far are shrub or bush roses. However, that other part of the rose family—the Climbing Roses—has also had a very important role in the development of English Roses.

The Noisette Roses were the first repeat-flowering Climbing Roses. They are usually tall and strong in growth, often capable of reaching 6m/20ft or more in height. Their flowers are generally in white, blush, pink and pale yellow shades. Their great virtues, other than their repeat-flowering qualities, are the delicate, silky beauty of their flowers and their ability to resist disease. When a climbing Noisette is hybridised with an English shrub, it will sometimes produce climbers, but more often than not will result in shrub roses. The Noisette Roses have brought a new character to a whole section of the English Roses, adding their grace, beauty, variety and interest—as well as a distinct fragrance. Most of these hybrids are short shrub roses; however, not surprisingly, there are among them a number of good climbers of the English type.

The most important group of climbing roses in the twentieth century was the climbing Hybrid Teas, most of which are climbing sports of the bush varieties. They make good climbers but, with the exception of 'New Dawn', have played no part in English Roses.

Rosa wichurana (syn. *R. wichuraiana*) is an extremely hardy, disease-resistant and vigorous climbing rose or, as gardeners would say, 'rambler'. Some of the best Rambler Roses have been bred by hybridisation with this species. They are summer-flowering only and do not repeat. This is not surprising for roses of such height and vigour. One of the best of the Wichurana Ramblers is a variety called 'Doctor W. Van Fleet'. By good fortune, in about the year 1930, this rose produced a repeat-flowering sport which was preserved and introduced under the appropriate name of 'New Dawn'. Although equally hardy and disease-resistant, 'New Dawn' is less strong than its parent; no doubt as a result of the effort of flowering throughout the summer. It has been widely hybridised with Hybrid Teas and other roses and has given rise—either directly or indirectly

—to a large number of good Climbing Roses. We used one of these, a variety called 'Aloha', for hybridising with our own varieties and produced a somewhat different group of English Roses that are closer to the Modern Rose in their foliage yet still with very much the Old Rose form of flower. Again, many of the hybrids were shrubs, although some turned out to be excellent climbers.

ABOVE 'New Dawn' (1930) was the founding variety of a race of Climbing Roses, which are very hardy and repeat-flowering.

4

THE
QUALITIES
OF THE
ENGLISH
ROSES

OPPOSITE A fine display of English Roses in our small garden in Wales, including 'Teasing Georgia' (*bottom, yellow*), 'Geoff Hamilton' (*centre right,* pink), 'Wildeve' (*centre,* pink) and 'Golden Celebration' (*top,* yellow).

THE ROSE in its numerous forms has many talents—although no group of roses has all of them. The Old Roses have beauty, charm, fragrance and pleasing, shrubby growth. The Modern Roses have a wide range of colour and the ability to flower throughout the summer. My aim has been, insofar as is possible, to combine all these virtues in one group of roses and, indeed, in one rose.

Every plant breeder has to have a clear idea of what he wants to achieve for his chosen flower. The first question I had to ask myself was: 'What should this be?' The second was: 'How can I achieve it?' I consider practical points, such as disease-resistance, freedom of flower and hardiness, to be as important in the development of English Roses as aesthetic factors, which may include broadening the range of flower colour, bringing in greater fragrance and so on. There is, however, one goal that I believe to be more important than all the others. This is, quite simply, that we should strive to develop the rose's beauty in flower, growth and leaf. This might seem an obvious objective, for do not all flower breeders have this aim in view? They may search for brighter and more intense colours; they might try to produce a larger flower and all manner of other measurable characteristics—but not many of them search for anything so abstract and elusive as the simple beauty of the flower and its growth which, of course, in the gardener's eyes, is the only reason for growing it. It seems to be assumed that all plants are beautiful whatever we do to them, and there is a certain truth in this. But it is quite possible, indeed almost usual, for the plant breeder to reduce this beauty while improving the more practical aspects of the plant. We need only look back down the ages to when man first started to adapt plants to his wishes, to see that in the end he nearly always degrades the very thing he loves.

Nature, left to her own devices, finds it hard to produce anything that is ugly. But the garden rose is not by any means entirely a product of nature. The hand of man has played a decisive part. The same is true of other highly developed flowers: dahlias, chrysanthemums, rhododendrons, daffodils, irises, lilies and peonies, to name but a few. In many cases, the work of the plant breeders has been beneficial and the subject has been enhanced as a garden plant. In many other instances, the effect has been damaging. We have only to walk around the average garden centre to see how true this is:

The Origins and Nature of an English Rose

giant, over-sized polyanthus and pansies, two plants that once had simple charm and beauty—and what have we done to them?

The result has been that many discerning gardeners (with somewhat highbrow inclinations) have preferred to confine their planting to flowers of nature, or at least to close relatives of species plants. I think this is a shame: the 'garden' flower can be developed to have its own kind of beauty, different from that of the simple child of nature. The garden is not the 'wild'; it is a man-made creation and should, with certain exceptions, be treated as such. The garden requires bolder statements than we would find in the wild.

With the above principles in mind, I describe below the various wishes I have for the English Roses, taking one aspect at a time, but always remembering that the breeder's skill is to capture the beauty of a rose at all stages. The crowning glory of the rose—its fragrance—is a characteristic I regard as being so important that I give it a separate chapter.

The Form of the Flower

The form of the flower is, perhaps, the most important feature in defining the rose's beauty and is also the aspect that differentiates more than any other the English Roses from other contemporary roses.

A quirk of nature allows the doubling up of petals in garden roses. What happens is this: the stamens of the flower mutate into petals over many years so that eventually, by means of selection, we have a 'double-flowered' rose. The wild species roses are undoubtedly beautiful, but it is the multiplication of their petals that has made possible the great wealth of beauty we now find in garden roses. It is by the reflection of the light between the petals and through the petals—and the many effects thereby produced—that all the beauty of the garden rose becomes possible. There are, in fact, no double roses in nature.

No variety of English Rose fits exactly into any one form but there are six basic shapes, with endless gradations, which come from the Old Roses, and a seventh which has the bud shape of Hybrid Teas.

SINGLE FLOWERS The rose is naturally single-flowered. In this form it charms us with its simplicity, delicacy and apparent frailty—to which must be added the airy beauty of its yellow, gold or red stamens. The beauty of the single rose, more than any other, depends upon the manner in which the flowers are poised upon the stem. Its growth should not be too heavy but light, open and graceful. We have only a few single-flowered English Roses: in fact, rather surprisingly, they are not the easiest to breed —perhaps because the garden rose has been double for too many generations; also, we have not many single varieties to breed from. Examples of single-flowered English Roses include 'Ann'—with particularly daintily poised flowers—and 'The Alexandra Rose' which forms a large, spreading shrub with typical wild-rose flowers. Single roses are generally not so fragrant as double roses, because they have fewer petals to make the fragrance. We have a number of very beautiful single-flowered English Roses in our trial grounds.

SEMI-DOUBLE FLOWERS   It is only a small step from the single to the semi-double flower: add a few petals and we have a rose which will last longer but is, in many ways, similar. Semi-double roses retain much of the light, dainty effect of a single rose and usually have attractive stamens. The flowers vary considerably: they may be cup-shaped or flat, formal or more loosely petalled; they may be produced in sprays of many flowers or few. They offer quite a large variety of rose beauty. Examples are 'Windflower', 'Scarborough Fair' and 'Cordelia'.

THE ROSETTE-SHAPED FLOWER   The rosette-shaped flower could be described as the quintessential English Rose. The rosette is also the main flower shape of the Old Roses. It is the intertwining of the petals that offers us the opportunity of a great variety of beauty. Such flowers may be either

loosely or closely packed with petals. Where there are fewer petals, it may be possible to detect just a few stamens. In some varieties the centre petals of the flower may be clustered together in the form of a button. Again, the petals may be incurved or flat—or they may reflex at the edges. In fact, a great variety of forms and effects become possible, so that it is almost impossible to find two flowers, between one variety and another, that are the same.

Varieties of English Rose with rosette-shaped flowers are too numerous to list in full but include 'Eglantyne', 'Mary Rose', 'Teasing Georgia' and 'The Countryman'.

THE DEEP CUP   Perhaps the most impressive flowers are those that are deeply cupped. These may be filled with petals, as in 'Brother Cadfael', or they may be, to a greater or lesser degree, open and goblet-like, as in 'Golden Celebration'. With the open-cupped flower, it is particularly pleasing to peer inside and see the intertwining petals—and perhaps sniff the flower's rich perfume.

Not many cup-shaped English Roses are completely open: often there are a few petals just hiding the stamens and in the full cup there are no stamens to be seen at all. Cup-shaped English Roses include some very large flowers that we might expect to be clumsy, but this is seldom the case. On a sufficiently large shrub they hold their heads gracefully on the branch which counteracts any impression of heaviness. Examples are 'Heritage', 'Jude the Obscure', 'Scepter'd Isle' and 'The Ingenious Mr Fairchild'.

THE SHALLOW CUP   Between the rosette and the deep cup lies the shallow cup, which can be equally attractive and often offers some of the most perfect of all flowers, the outer, incurved petals forming a kind of frame for the numerous petals within. 'Crown Princess Margareta' and 'Teasing Georgia' are examples.

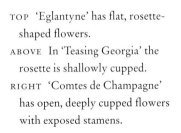

TOP 'Eglantyne' has flat, rosette-
 shaped flowers.
ABOVE In 'Teasing Georgia' the
 rosette is shallowly cupped.
RIGHT 'Comtes de Champagne'
 has open, deeply cupped flowers
 with exposed stamens.

OPPOSITE 'Golden Celebration'
 has giant, cupped blooms filled
 with petals.

The Origins and Nature of an English Rose

BELOW 'Marinette' has dainty, long
   thin buds and opens to a semi-
   double flower.
BELOW RIGHT 'Janet' has pointed
   buds at first, which are somewhat
   similar to those of a Hybrid Tea.

THE RECURVED FLOWER   A number of English Roses have blooms in which the petals reflex to form a recurved or dome-shaped flower. This type of flower often starts its life as a rosette or even as a shallow cup. It then develops into a flat flower and finally the petals turn back—sometimes to the point where they almost form a ball. The flowers of all double roses have the advantage that they are always changing. This is particularly true of the recurved flower—it is as though we have a new bloom at each stage. Examples are 'Grace' and 'Jubilee Celebration'.

THE BUD FLOWER   English Roses were seen as something of a revolution when they first appeared in the 1960s—which, at that time, they were—though they were in fact a return to what the garden rose had always been. But what of the Hybrid Tea, with its long, pointed bud? While the growth of the Hybrid Tea, as we know it at present, tends to be short and almost entirely lacking in grace, I am strongly in favour of the Hybrid Tea bud, so long as it eventually develops into a beautiful open flower and is held nicely upon attractive, shrubby growth. Indeed, we already have such roses

among the English Roses. The variety 'Janet', for example, often has attractive Hybrid Tea buds which open to a rosette-shaped flower, both bud and open flower hanging elegantly on quite long branches on a sizeable shrub.

Texture, Light and Size of Flower

It is not simply the shape of the flower in an English Rose that sets it apart from other modern roses; the texture of the petals and the manner in which the petals interplay within the flower also contribute much to its beauty. A variation in the texture of the petals inevitably changes the whole character of a flower. Most English Roses are softer and more transparent of petal than other contemporary roses. This enables them to produce a more gentle, glowing, ever-changing effect as the light reflects through and between the petals.

There is also the question of the way in which the bloom of a rose is put together. Some are strictly formal, while others are more loosely arranged. The petals within some flowers are displayed in neat swirls as they progress towards the centre, whereas others twine and intertwine in the most fascinating manner. All forms have their beauty, even the apparently random arrangement of petals on loosely formed flowers having an appeal.

Size of flower is not important—a rose can be beautiful in all sizes. There is no doubt that a large flower produces a dramatic effect, but if all roses were large, this would be lost. I am always sorry when I see one of our customers choosing a rose simply for its size and brilliance of colour. Each size has its own particular beauty and variation in size offers us the chance to bring diversity to roses.

ABOVE LEFT 'James Galway' has very full, domed flowers, the outer petals gradually recurving a little.
ABOVE RIGHT 'Charles Rennie Mackintosh' has the soft texture of many English Roses.

BELOW, TOP 'Gertrude Jekyll',
a deep pink English Rose.
BOTTOM 'The Generous Gardener'
is a most delicate blush pink.

Colour

If form is the most important characteristic of the flowers of a rose, colour is not far behind. Here we are looking not simply for variety of colour, but for good colours and those that are suitable for the rose. A colour that is entirely appropriate for an iris may be less than ideal for a rose—particularly an English Rose. Dress up the rose in gaudy colours and she sometimes does not look right.

SHADES OF PINK   Above all others, pink is the colour of roses and most wild species are pink. When I began breeding the English Roses in the early 1950s, *pure* rose pink had almost disappeared from the Hybrid Teas, the pink nearly always being mixed with other colours. These were not necessarily bad colours, although some of them were rather muddy. Consequently, my first aim was to produce roses with flowers of purest glowing pink—the pink that we find in the Damask Roses and the Centifolias.

'Constance Spry', the first variety introduced by David Austin Roses, is a fine example of a glowing pure rose pink. Others followed, such as 'Cottage Rose', 'Gertrude Jekyll' and 'Sharifa Asma'. Soon we started to produce English Roses of delicate blush pinks, often with a Noisette Rose

background, such as 'Eglantyne', 'Heritage' and 'Perdita'. More recently we have bred roses in blush shades tinged with apricot, peach and orange. Indeed, there is no colour in roses that offers us the possibility of so much variation as do the various shades of pink. Put a bunch of red roses together and they tend to look rather similar. Much the same could be said of the stronger yellows. This may be because these two colours are often very intense and do not always mix easily with other shades. But the many shades of pink among roses seem to be unlimited and each shade is surprisingly well defined. Pink roses—particularly soft pinks and blushes—are often used as symbols of femininity and childhood. They have a gentleness that we do not find in other colours.

CRIMSON AND OTHER SHADES OF RED   A deep crimson rose is so special that it seems to be a flower apart from roses of other colours. Red is a symbol of passion and is often the most popular colour with men. However, crimsons and other reds tend to be something of a challenge for the rose breeder. There are almost no deep red roses in nature, except for the beautiful species *Rosa moyesii*. Examples of this rose found in gardens usually have flowers of the clearest deep crimson imaginable, but these are the result of selection. In the wild they are normally deep pink in colour. The colour crimson came to the garden rose via a China Rose, probably 'Slater's Crimson China'. This is a rather puny variety and it seems to have handed its weakness down to its descendants. In spite of this, there are some red Floribundas that are both strong and healthy—and also, more recently, a few Hybrid Tea Roses.

We have gone to great lengths to produce good, clear crimson varieties, but still they are not plentiful among the English Roses. However, 'L.D. Braithwaite' is a good example of bright crimson colouring, while 'Benjamin Britten' is closer to scarlet.

LILAC, PURPLE AND MAUVE   These shades are some of the most evocative, having associations with royalty, majesty and power, as well as with emotions allied to grief and melancholy. The old Gallica Roses were notable for lilac, purple and mauve and, since English Roses have elements of Gallica in their breeding, these colours show up again here. We also find similar colours in Modern Hybrid Teas and Floribundas, but these tend to have a rather metallic appearance, which is much less pleasing. Such shades in the Gallicas possess a depth and a richness that we do not find elsewhere and these qualities have been passed down to the English Roses.

English Roses in rich shades of purple, mauve and lilac usually start their life as deep crimson and it is in the progressive changes of colour that we get so many beautiful tints. An example is 'The Prince', which starts off as a deep, rich crimson but quickly turns to a rich shade of royal purple. Unfortunately, this rose is not very robust. 'William Shakespeare 2000'—possibly

TOP 'Gentle Hermione' is soft pink at the centre, shading with age almost to white on the outer petals.
ABOVE 'Falstaff' is a strong-growing crimson rose that leans towards the Bourbon Roses in character.

The Qualities of the English Roses   37

BELOW 'William Shakespeare 2000' is a magnificent large, full rose of dusky crimson.

the best of our crimson roses—also starts off as a deep, pure crimson but passes through all manner of shades of intermingled purple and mauve, providing the most beautiful effects. It has excellent vigour and a rich fragrance.

YELLOW, APRICOT AND PEACH   English Roses in shades of yellow, apricot and peach are so unlike those in the range from blush pink through to deep pink and on to crimson and purple that they almost seem to belong to another species. In fact, the origins of yellow roses in the wild are very different.

Certainly you can have a beautiful border or even a rose garden made up entirely of yellows, just as you can have one of pink shades. The yellow shades seem to denote happiness and sunshine. Most of the yellow English

Roses belong to the Leander Group, although a number of soft yellows are to be found in the Musk Group. The soft yellows seem to make better companions for other roses and other plants than do the stronger yellows, although, of course, all these shades have their place in the garden. Soft yellows include 'Pegasus' and 'The Pilgrim'. Among deeper yellows, 'Golden Celebration' and 'Teasing Georgia' are particularly fine, while 'Graham Thomas' is one of the richest, purest yellows I know, and 'Grace' is of a lovely apricot shade.

SHADES OF COPPER AND FLAME   Because the Hybrid Tea has a tendency towards gaudiness, we have concentrated on breeding the English Roses with softer colours. When we do give attention to brighter colours, we try to be sure that these are in shades suitable for a rose. It is not easy to breed roses in these shades and they tend to come more by chance than by design. A flame-coloured rose is seldom bred from two varieties of the same colour; it is more often the result of a cross between a red rose and a pink. It is important that such colours should not be simply another colour, but a good one. 'Pat Austin' is an example of a rose of a lovely copper colouring.

WHITE   With the possible exception of the old Albas, there are very few really good white varieties in any class of roses—and this is particularly true of English Roses. Not many breeders actually set out to breed white roses. They are often the product of a breeding programme with other aims in view.

Nonetheless, white is a very important colour—if we can call it a colour. There is a lily-like freshness about white roses, something quite different about them that is hard to explain—they are all purity and form. A few people have gone so far as to make a rose garden entirely of white roses to very

TOP LEFT 'Grace' is a beautifully formed rose in a lovely shade of apricot.

TOP RIGHT 'The Pilgrim' is a good example of an English Rose in a soft shade of yellow.

ABOVE 'Pat Austin' has the brightest and most lively colour of all the English Roses.

good effect, although not many of us could spare the space for such a garden. In any case, this would not be possible with the English Roses, with their very limited range. They include 'Winchester Cathedral' (a sport from 'Mary Rose') and 'Crocus Rose', the latter not a pure white, as it occasionally has tints of palest apricot. Indeed, one problem with breeding white roses is that they are often not quite pure and have tints of other colours in their make-up. There is also 'Glamis Castle' which, though excellent in other ways, tends to suffer from blackspot. 'Francine Austin' is an extremely good white rose, but it is a spray-flowered variety and therefore not quite a typical English Rose. We are now developing strains with a heavy bias towards white and hope to have some interesting roses in the not too distant future.

Growth and Foliage

There is a tendency to think of roses in terms of the flower alone, overlooking the rose as a complete shrub with its own beauty, which makes it a valuable garden plant. This, at least, is true of English Roses and many other shrub roses. Indeed, up to the second half of the nineteenth century, roses were excellent garden shrubs. It was only with the arrival of the repeat-flowering roses from the Far East, and their hybridisation with the old summer-flowering roses, that the rose began to lose its way as a garden plant.

The original European roses, which we now know as the Old Roses, flower once in early summer and thereafter put out strong shoots for the production of the next year's flowers. The new roses produce a flower on every shoot and, if every shoot produces a flower, repeat-flowering results. This leads to a somewhat ungainly plant, although the repeat-flowering shrubs of the twentieth century did correct the fault to some extent. English Roses are, in part, an attempt to produce a better garden shrub while retaining its repeating-flowering qualities.

The English Roses, unlike the Hybrid Teas and Floribundas, have natural, shrubby growth. They may be bushy and twiggy; they may be tall and upright, making them suitable for planting behind other subjects; or they

OPPOSITE Although English Roses are by nature shrubby, many of them can be pruned hard to form short shrubs or bedding roses.

BELOW 'William Shakespeare 2000' has nice spreading growth.

may have long, elegant, arching growth. Their height can vary from 1m/3ft to 2.4m/8ft or more. They may be broad or they may be narrow. Their growth may be open or it may be full. All these factors give us a great range of growth suitable for a variety of positions in the garden.

Just as the growth of English Roses varies, so too does the foliage. This is not surprising when we take into account the many parents from which they sprang. The foliage may be fine and species-like or broad and heavy, or it may be anything between these extremes. Its colour may be pale green or of a greyish, almost glaucous nature, or it may be in all the gradations of green up to an almost black-green. Thus we have a great variety of foliage for a great variety of flowers. English Roses—and indeed all kinds of roses—are worthwhile subjects for their foliage alone. This is never more evident to me than when I walk around our gardens in spring and early summer, before the flowers appear.

Growth and foliage are not only important in themselves, but they also make an important contribution to the beauty of the flower. The way in which the flower is held upon the shrub and the foliage against which it is set provide us with something of greater beauty than the bloom alone. We increasingly try to exploit this fact in our breeding, but there is much more that can be done.

Not only do growth and foliage have a beneficial effect on the bloom and the plant as a whole; they are also important in the use of English Roses in the garden. Different growth is suitable for different positions in the garden and in association with other plants. Moreover, if you have a dedicated rose

The Origins and Nature of an English Rose

border or a rose garden, it will benefit greatly from all this variety. A border of Hybrid Teas and Floribundas does not lack beauty of bloom, but the sameness of growth and foliage is boring.

I cannot over-emphasise the importance of growth and foliage in the development of the English Roses and of suiting the qualities of these to the flowers. 'Golden Celebration' is a good example of what can be done. It has very large, heavy flowers of cupped formation which could look clumsy were it not for the fact that they hang gracefully upon the branch. 'Mary Rose' and 'Eglantyne' are examples of twiggy, bushy growth. 'The Countryman' and 'William Shakespeare 2000' are low and spreading, while 'Windflower' and 'Cordelia' are much closer to the wild rose in their growth. All in all, we have developed roses with a high degree of beauty both in flower and growth, but there are still great possibilities for the future.

Some Practical Considerations

Breeding roses is a complex and difficult business. There are so many different characteristics in which the rose has to excel if it is to achieve the beauty we require of it—and it must excel in each one of them. This, however, is not all. There are also practical attributes that a rose must possess, at least to a sufficient degree, if it is to be suitable for the garden. I believe that if we cannot make the rose a more healthy and vigorous plant, it does not have a very bright future.

DISEASE RESISTANCE   Gardeners of the past—perhaps with more time on their hands than we have today—were happy to spray their roses. It was as though they enjoyed the extra challenge. Many keen gardeners still feel this way but, on the whole, modern gardeners prefer not to spray. There is also the fact that some people do not like chemical sprays and believe them to be dangerous—perhaps a little unreasonably, since we are not going to eat our roses! But, in any case, it is better to have roses that do not require spraying, first of all because it saves work, but also because they will always appear fresher and grow better than those we do have to spray.

Fortunately much work is being done on developing disease-resistant roses, some of it by professional rose breeders and research institutions—particularly in the USA and Canada—but much good work has also been done by amateur breeders. We ourselves have benefited greatly from these efforts in the course of breeding the English Roses. Most experts agree that English Roses are among the most disease-free of all garden roses. This is largely due to the fact that we—so to speak—started again, using parents that had a high degree of vigour and disease-resistance. Some of our varieties are, as far as we know, *entirely* free of disease. We have sent bushes of these roses, for trial, to a large number of expert gardeners around the world and

ABOVE 'Cordelia' is a very healthy rose of robust growth.

OPPOSITE 'The Mayflower', a charming English Rose, is unique in that it is completely free of disease, insofar as we know.

The Origins and Nature of an English Rose

they have proved to be virtually 100 per cent disease-free, although occasion-ally they will become slightly infected just as their leaves fall in the autumn. It is impossible to be certain whether this is true in all climates and in all countries. We hope to have many more such healthy roses as time goes by.

VIGOUR AND FLOWERING   Vigour and the ability to flower—at least intermittently—throughout the summer are further practical factors that have to be taken into account. We are not all of us experienced garden-ers, nor are we always blessed with the best of soil. It is obviously desirable that a rose should be vigorous and if possible thrive, even in less-than-per-fect conditions. This will usually mean that they are easily grown, even with sometimes less-than-ideal cultural attention.

Closely allied to vigour is the ability of a rose to repeat-flower regularly. Up to the time of the introduction of the English Roses, there were not many Shrub Roses of the old type that could be relied upon to repeat-flower with any degree of regularity, and even when they did, they were seldom good roses in other respects. All English Roses, with the exception of 'Constance Spry', 'Chianti' and 'Shropshire Lass', will provide us with at least two flowering periods—the first early in the summer and the second in late summer, with a good smattering of bloom between, so that they are seldom out of flower. Many varieties will provide an even more continuous display, especially in countries with long, warm summers. 'The Mayflower' is particularly reliable in this respect.

The Art of Rose Breeding

The various aspects of the breeding of English Roses—from the form and colour of the flower to the plant's vigour and disease-resistance—may be worth taking into account when choosing roses for your garden. There is, however, one more point, and it the most important of all. It is an extra *something* that is hard to explain, which I described as 'beauty in flower, growth and leaf' at the beginning of this chapter. As well as 'beauty', we use such words as 'charm', 'character', 'freshness' and 'elegant simplicity' as we

cast around for ways to communicate a quality that we know when we see, although each of us has our own tastes and preferences. There is an excellence in certain roses that is unmistakable. This has a way of coming through in the breeding. In the English Roses we do our best to take advantage of this fact and we constantly endeavour to develop strains of even greater beauty.

As we walk through our fields of trial seedlings, observing and noting their attributes, it is this very special quality we look for above all others. The flower of a certain seedling can have all the virtues outlined in this chapter and still not be quite what we desire. The special quality for which we are searching is probably something to do with the way in which the flowers catch the light. It is something one knows when one sees it, particularly when it is combined with the all-important factors of growth and foliage and the manner in which the bloom presents itself on the shrub.

The word we use on the Nursery is 'charm'. In fact, we use the word to measure the value of one seedling rose against another, recording it on a scale of 1 to 10.

The Origins and Nature of an English Rose

Although the best roses become known as time goes by and are hard to forget, it is not possible to remember thousands of different seedlings when selecting in our trial grounds and this measurement enables us to return to a seedling—as we do again and again—with some notion of what we felt about it at other times of the year under other weather conditions. Fortunately, we have all kinds of weather in the British Isles, which is perhaps why our country is regarded as one of the best places to breed roses. If a rose does well here, it will probably do well in most climates.

Of course, when we have both flower that charms and growth that appeals to us, there remain the important questions of health, vigour and repeat-flowering. More times than I care to remember we have found a rose that delights us with its beauty, only to have to condemn it because of some weakness in these areas. This slows down the breeding and selection process enormously. If we could conquer these problems, the breeding of roses would be a very much easier task.

5

FRAGRANCE

FRAGRANCE may be said to be the other half of the beauty of a rose and to bring greater fragrance back to the rose was one of the main objectives in the breeding of English Roses, and the one in which we have been particularly successful. I think most gardeners will agree that, if nothing else, English Roses are almost all very fragrant; more so than any other group.

A sense of smell is a very private affair and has a limited use. It can warn us against danger, it can delight or disgust us; but there is no art of smell—although it can be a great source of pleasure. One of the very few ways in which the sense of smell becomes something like an art is through the fragrance of flowers. And among these flowers, roses provide us with the most complex and the most beautiful range of scents. The distillation of rose essences has been a flourishing industry in certain parts of the world for centuries. Rose water has long been believed to have curative properties.

Perhaps more importantly, the fragrance of the rose has been widely used in perfume for personal use and is still the basis of most manufactured perfumes.

The scent of the rose tends to be an elusive quality—indeed, this adds to its charm. We never quite know what we are going to get. So many varying elements come into play. A rose can be fragrant today and less so tomorrow—often because weather conditions vary so much. Warm, moist weather usually results in greater fragrance. Not only the strength but the kind of fragrance is affected by weather, certain conditions bringing out certain fragrances—although, in fact, it is not the weather on the day we sniff the rose so much as what it was like on the day before that influences its fragrance. It is at that time the fragrance is triggered. Then there is the fact that roses smell a little different according to the age of the flower. It seems that certain chemical ingredients are used up more quickly than others as the flower ages. A rose is most fragrant in the

INSET 'Jude the Obscure' has a strong fragrance with a fruity note reminiscent of sweet white wine.

OPPOSITE 'Gertrude Jekyll' has a very strong but well-balanced Old Rose fragrance which is one of the most pleasing in English Roses.

The Origins and Nature of an English Rose

BELOW 'L.D. Braithwaite' is
interesting in that it has only
a light scent until the flower ages,
when it develops a charming
Old Rose fragrance.

morning, perhaps because it has had time to gather strength during the night. As the day lengthens, certain chemical constituents of the flower's fragrance decline and we get a rather different perfume. The season, also, has an effect on fragrance: we might have one scent in early summer and quite a different one in late summer. Usually the fragrance is better early in the year—although it is interesting to note that the rose 'L.D. Braithwaite' has very little scent early in the season and a good scent later on. Geography, too, can have a marked effect. A rose may have little scent in Britain and an almost overwhelming fragrance in warmer climates like those of Australia and the southern USA. All these variations are no doubt connected with soil and climate and all add to the fascination of the fragrance of the rose.

We ourselves play a part in these many differences. The fragrance of the rose is made up of numerous chemicals, and some people are able to detect most or all of these, while others detect only certain of them and therefore enjoy a rather different fragrance. Much depends on our ability to pick up particular 'notes'.

Rose breeders, over the years, have tended to ignore fragrance in their breeding programmes. This is because people seldom buy a rose for its fragrance; they nearly always have the appearance of the flower in mind when they are making a choice. Most people regard scent as an added extra, although when the rose is growing in their garden, they are nearly always dissatisfied if they find that there is little fragrance. This is probably the reason for a decline in the fragrance of roses over the last hundred years. Many breeders see no commercial advantage in good and strong scents.

We cannot claim to have consciously bred the English Roses for certain shades of scent, although I can foresee a time when this might be done. However, we have drawn upon nearly all the groups of roses that have come down to us from the past and done our best to choose such parents as have very strong fragrances and—insofar as is possible—good fragrances. We select from their progeny those that have the most beautiful scent. As a result, English Roses are not only the most fragrant group of roses, but they also have by far the widest range of fragrances.

Nearly all the basic scents of the rose are to be found somewhere among English Roses and, as a rose of one scent is hybridised with a rose of another, new scent combinations become evident. So it is that we find one scent merging into another, as we move through the varieties of English

Roses. I regard this as one of the greatest pleasures they have to offer us.

One problem arising out of this great diversity of fragrances in English Roses is the difficulty in describing them. It is rather like writing about wines; in fact, taste is, as we all know, very close to the sense of smell. We can but do our best, by means of classification and reference to other scents that most of us know. As with wines, there is the danger of sounding pretentious.

We have been fortunate, in recent years, to have had the benefit of the expertise of Mr Robert Calkin, consultant perfumer and a great lover of roses.

Fragrance, like the visual beauties of the rose, has its own art, as we mingle one scent with another. While the scent of roses varies widely from one variety to another and from one condition to another, there are certain basic fragrances of English Roses.

BELOW 'Rosemoor' is a dainty little rose which has an Old Rose fragrance with hints of apple, cucumber and violet leaf.

The Old Rose Fragrance

The so-called 'Old Rose' fragrance is to me the most beautiful of the many fragrances of English Roses or, indeed, of any roses. The majority of the Old Roses have this fragrance. We find it first of all in the Gallicas and again in the Damasks and Centifolias, as well as in many of the Alba Roses. Later on, it appears in the Portland Roses and often in the Hybrid Perpetuals. Rather surprisingly, the same fragrance is also found in *Rosa rugosa*, a strong-growing species found wild on the coasts of Japan and Korea. The Old Rose fragrance, like that of all roses, is seldom quite the same in any two roses. There are many variations and, as my descriptions show, this is particularly true in the English Roses, which are of more mixed origin.

Among English Roses, 'Gertrude Jekyll' has the finest and possibly the strongest Old Rose fragrance. This is inherited from its pollen parent, the Portland Rose 'Comte de Chambord'. 'Eglantyne' has an Old Rose fragrance of a lighter character. This is inherited from a Rugosa parent earlier in its ancestry. Another particularly fine example of an Old Rose fragrance is that of 'Noble Antony'. At the Glasgow Trials it was described as a 'glorious Old Rose perfume that has a hint of the "oaky" character, such as we find in the Gallicas and similar to that of a well-matured red wine'. Both 'Brother Cadfael' and 'Sir Edward Elgar' have an excellent Old Rose fragrance, but require warm weather to give of their best. 'James Galway', named after the flautist, also has a lovely Old Rose fragrance. 'The Countryman' is

OPPOSITE
TOP 'St. Cecilia' has a myrrh fragrance
mixed with lemon and almond blossom.
BOTTOM 'Blythe Spirit' has a light Musk
Rose perfume with a hint of myrrh.

BELOW 'Scepter'd Isle' has the rare
myrrh fragrance that is found
only in English Roses and one or
two Ramblers.

a close relative of 'Gertrude Jekyll' (indeed, both roses were discovered close together in our trial grounds) and has a fine Old Rose fragrance inherited from 'Comte de Chambord', but this is accompanied by a fresh strawberry character that comes from a seedling descended from the Climbing Rose 'Aloha', which has *Rosa wichurana* connections. All these roses have passed the Old Rose fragrance on to later introductions.

The Tea Rose Scent

This lovely, fresh fragrance arose from the hybridisation of the China Roses with *Rosa gigantea*, a tall Climbing Rose that bears large, single, white flowers, and was so named because its scent closely resembles that of a freshly opened packet of China tea.

The Tea Rose scent is usually found in English Roses in shades of yellow and apricot. Among the varieties notable for this scent are 'William Morris', 'Pegasus' and 'Molineux', the last having been awarded the Royal National Rose Society's Henry Edland Medal for Fragrance. The rich yellow rose 'Graham Thomas' provides us with a particularly fine example of an English Rose with a prominent Tea Rose fragrance. The brightly coloured 'Pat Austin' also has a strong Tea Rose scent. The Tea Rose fragrance has had an important influence on other fragrances found among the English Roses, combining effectively with myrrh fragrance—as in 'The Pilgrim'—and with the fruity scents, as in 'Crown Princess Margareta'.

Myrrh Fragrance

Myrrh fragrance is almost unique to English Roses. As far as I am aware, the only other garden roses in which it occurs are the Ayrshire Roses—in fact, the variety 'Ayrshire Splendens' (now named *R*. 'Splendens') is also known as 'The Myrrh Scented Rose'. The Ayrshire Roses are some of the oldest ramblers we have.

The myrrh note was introduced with our first English Rose, 'Constance Spry', the result of a cross between a Gallica called 'Belle Isis' and the Floribunda 'Dainty Maid'. 'Belle Isis' has quite small blooms of soft flesh pink. It is obviously not a pure Gallica and Graham Thomas suggested that it must be the result of a cross between an Ayrshire Rose and a Gallica Rose. Whatever the truth may be, it was 'Belle Isis' that first brought the myrrh scent to the English Roses and it has proved to be remarkably persistent down the generations.

The Origins and Nature of an English Rose

The myrrh scent is closely related to sweet anise. The fragrance of the rose 'Iceberg' contains a low level of anise in its make-up and it was an important early parent of certain English Roses. Its descendants may have further established the myrrh fragrance among English Roses.

Not everyone loves the myrrh scent, though most people find it pleasing. When, however, it is mixed with the Old Rose fragrances and others, the results are excellent. In the rose 'Constance Spry' there is an underlying Old Rose fragrance. In 'Heritage' myrrh is shaded with overtones of honey and fruit, together with a musky clove note.

The myrrh fragrance is found in a number of English Roses that are predominantly 'Tea' in character. Outstanding examples of this combination are present in 'The Pilgrim', in 'St. Cecilia'—where there is also a mingling of myrrh with Old Rose and Tea fragrances—and in 'Perdita' which was also awarded the Henry Edland Medal for Fragrance, although here the blend is powerful but a little less successful.

Musk Fragrance

There is something romantic about the idea of Musk fragrance in a rose. It owes its name to the fact that it is in some degree reminiscent of true musk—a secretion of the Himalayan musk deer, once widely used in perfumery. Unlike other rose scents, which occur in the petals, the Musk scent is produced solely in the stamens of the flower. As, over the generations, the rose has gained more and more petals by means of selection and become more double-flowered, so the stamens have gradually decreased—and the Musk scent with them. In roses where, by happy chance, the scent of the stamens is nicely balanced with that of the petals, we can get some very pleasing results. The old Gallica Rose 'Officinalis' (now *Rosa gallica* var. *officinalis*), for example, has a Musk Rose fragrance from its stamens, which blends beautifully with the Old Rose fragrance of its petals.

Generally, the scent of a rose does not carry far on the air and, to enjoy it to the full, it is usually necessary to sniff the flower at close

OPPOSITE

LEFT 'Heather Austin' has an
    Old Rose fragrance that combines
    with the scent of cloves as the
    flower opens.

RIGHT 'Barbara Austin' has a strong
    fragrance which is a mixture of
    Old Rose and lilac.

BELOW 'Abraham Darby' has
    a strong, rich fruity fragrance
    with a refreshing sharpness.

quarters. For this reason I would never recommend planting with roses such flowers as pinks (*Dianthus*), whose fragrance carries on the air and would overwhelm the fragrance of the roses. The Musk fragrance, however, is different—its scent will fill the air around the rose and we can enjoy it as we pass.

A number of English Roses have the Musk fragrance, although it usually comes with other fragrances and may not be dominant. We find it in 'Blythe Spirit', 'Comte de Champagne', 'Francine Austin', 'Molineux', 'The Generous Gardener', 'Windrush' and others.

Fruit Fragrances

A wide range of fragrances are associated with fruit, although we seldom notice these while enjoying the taste of fruits. That roses should have the fragrance of fruit is not so surprising when we consider that the majority of the fruits we eat, such as apples, pears, raspberries, strawberries, apricots, peaches and so on, are members of the same botanical family as the rose—*Rosaceae*—although we do also get the scent of guava (*Myrtaceae*) and lemon (*Rutaceae*).

Of course, when we talk of a 'fruit-like' fragrance, we mean a typical rose fragrance with fruit-like undertones or, perhaps, overtones. However, while fruit fragrances are notable for their influence on other scents rather than as fragrances in their own right, they are nonetheless easily recognisable. We find hints of fruity scents in the China Roses, the Autumn Damasks and their descendants, the Bourbons, and the Hybrid Perpetuals—all of which have some connection with the breeding of the English Roses. *Rosa wichurana* carries a pronounced apple scent and this finds its way into the English Roses by way of the Climbing Rose 'Aloha', which has had considerable influence on many of our stronger-growing roses of the Leander Group, such as 'Abraham Darby', 'Golden Celebration' and 'Crown Princess Margareta'. Gradually the *Rosa wichurana* fragrance has become mingled with other fragrances.

The Origins and Nature of an English Rose

'Golden Celebration' also has a sharp lemon Tea character, combined with its fruity fragrance. 'Jude the Obscure' has one of the most fruit-like fragrances of all the English Roses. When fruit-like fragrances are intermingled with those of other roses, they produce a great variety of rich and lovely scents. Indeed, they have enhanced the fragrance of English Roses considerably.

Myriad Fragrances

As well as the basic fragrances of English Roses there are many others, but they usually come mixed with the fragrances outlined above. Sometimes it seems as though the fragrances of all the flowers are to be found somewhere in English Roses. The scent of lilac is found in 'Heather Austin' and 'Barbara Austin'; that of lily of the valley in 'Miss Alice'. The scent of peach blossom is found in a number of roses. Sometimes, as we cast hither and thither for a name for our fragrances, we refer to them in terms of the bouquet of wine or the fragrance of honey. Clove scent occurs in certain varieties, as, for example, in 'Heritage'. Seldom are these comparisons exact. Not always can any two people agree on the right term, but this only adds to the many charms of English Roses.

6

THE FIRST ENGLISH ROSES

I HAVE DESCRIBED how my idea for the English Roses was to combine the beauty and the shrubby growth of the Old Roses with the repeat-flowering abilities of the Modern Roses—together with their wider range of colour—and, in the process, produce a rose superior to both. A brief outline of my first steps in the creation of the English Roses will, I hope, give a clearer idea of my aims and objectives.

Most readers will know that new roses are usually produced from seed. One variety is cross-fertilised with another, in the hope of mingling the two sets of desirable characteristics and producing a rose with a combination of both. When I began breeding roses in the 1950s, few Old Roses were available and I had a comparatively limited collection of Old and Modern Roses from which to choose the parents. I selected, among others, the Gallica Rose 'Belle Isis' (bred by Parmentier of Belgium and introduced in 1845), for the delicate charm of its well-formed Old Rose flowers and also for its sweet scent and general good health. On the Modern Rose side I chose the Floribunda 'Dainty Maid' (bred by E. B. LeGrice in 1938); it was both strong and healthy and its flowers were large, single and of a good, clear shade of pink. This last point was important to me, since such clarity of colour was rare among the roses of that period. It was, as we might expect, also reliably repeat-flowering. My idea was to produce a small, shrubby plant of about the size of a typical Gallica Rose. I sowed a few hundred seeds from this

RIGHT Floribunda 'Dainty Maid' (*left*) and Gallica Rose 'Belle Isis' (*right*) were the parents for our first English Rose 'Constance Spry'.

OPPOSITE 'Constance Spry', with large, glowing pink flowers, is an excellent rose both as a strong-growing shrub and as a climber. Unfortunately it is only early-summer-flowering.

The Origins and Nature of an English Rose

cross, which resulted in a small batch of seedlings. When, eventually, the seedlings flowered, I was in for a surprise. The best seedling was not what I had expected. It was a large, sprawling shrub with gigantic, cup-shaped 'Old Rose' flowers of a glowing, clear rose pink. In spite of the size of its flowers, it was not ungainly, all parts of the shrub being in proportion. The flowers had a delicacy and refinement that it has been hard to improve upon in subsequent years.

I was disappointed to find that none of the roses in this batch of seedlings was repeat-flowering. Today I would have realised that this was to be expected, but at that time my knowledge was very limited. In fact, the repeat-flowering gene is nearly always recessive when a repeat-flowering rose is crossed with a non repeat-flowering one.

I took blooms of this outstanding variety, together with a number of others, and showed them to my friend and expert rosarian, the late Graham Thomas. He was most enthusiastic, a fact that I found encouraging. He immediately agreed to introduce this rose through Sunningdale Nurseries, of which he was then nursery manager, and it became available in 1961. I could not do this myself as I was, at that time, a farmer with no connection with the nursery business. We named the new rose 'Constance Spry' after the well-known flower arranger and it was an instant success. Indeed, with its gorgeous, rose-pink, peony-sized flowers, it remains to this day one of our more popular roses. To be presented with such a rose at this early stage was a great piece of beginners' luck—although not quite all luck, as it had just the kind of flower that I was aiming for. However, it has to be remembered that it is much easier to breed a good rose that is only early-summer-flowering than one that is repeat-flowering. We introduced 'Constance Spry' as a Shrub Rose, but in the course of time it proved to be even better when grown as a climber.

The fragrance of 'Constance Spry' is interesting. It has, as I have said, a strong myrrh scent that was unusual in garden roses, although now quite common among English Roses. Graham's suggestion that its parent 'Belle Isis' was itself a hybrid between a Gallica Rose and an Ayrshire Rose would explain the scent and also why 'Constance Spry' performs so well as a climber. The Ayrshires form a small group of Rambler Roses that bear small, usually white flowers, in sprays. They are the only garden roses, as far as I know, to have the myrrh fragrance. It has proved persistent among English Roses and we find it appearing again and again down the generations.

Another parent we used in those early days was the lovely, rich crimson Gallica Rose 'Tuscany Superb' (1850). It is not quite a pure crimson, but as pure as anything available before the arrival of 'Slater's Crimson China'. We crossed 'Tuscany Superb' with another LeGrice Floribunda called 'Dusky Maiden'. My objective was, once again, to produce a rose with flowers of

The Origins and Nature of an English Rose

truly Old Rose character, but this time with deep, rich crimson colouring. The result was a fine Shrub Rose of some 1.8–2.4m/6–8ft in height, with flowers of a pure crimson colouring that later changes to a rich purple. This rose had a powerful Old Rose fragrance. But, again, being a cross between a repeat-flowering and a non-repeat-flowering rose, it was itself not repeat-flowering. Its flowers were not quite so large as those of 'Constance Spry'; nonetheless they had considerable beauty and it was perhaps a better shrub in terms of growth habit. We named it 'Chianti', at Graham Thomas's suggestion, and, again, introduced it through Sunningdale Nurseries, in 1967.

We made numerous other crosses between Old Roses and Modern Roses, many of which led us down blind alleys, but 'Constance Spry' and 'Chianti' gave the most promising results, and accordingly it was upon them that we first based the English Roses. They both had the particular character of flower that we wanted, though they were still only summer-flowering. To obtain the repeat-flowering character, we had to back-cross again to a repeat-flowering variety of Modern Rose. This we did and one or two repeat-flowering seedlings began to appear. By the third generation of back-crosses, nearly all our seedlings were repeat-flowering and—I am pleased to say—had retained much of the Old Rose character which we so much desired.

Subsequently, many other parents were used, including the Gallica Roses 'Duchesse d'Angoulême', 'Duchesse de Montebello' and *Rosa* 'Officinalis' (now *R. gallica* var. *officinalis*); the Damask Roses 'La Ville de Bruxelles', 'Marie Louise' and 'Celsiana'; and also the Albas 'Königin von Dänemark' (Queen of Denmark) and 'Madame Legras de St. Germain' on the Old Rose side. On the Modern Rose side, we tried to select early varieties that were themselves not too far removed from the Old Roses in appearance, so as not to dilute the Old Rose effect. These included 'Madame Caroline Testout' (bred by Chauvry in 1901), which arrived on the garden scene before the Hybrid Teas had become—so to speak—too 'Hybrid Tea' in flower. It had very full, rather cup-shaped flowers of lilac pink. It was a very healthy and reliable rose which was widely grown in its day and might be described as the 'Peace' of that period (this rose is now named 'Madame A. Meilland'). Another rose used was the Floribunda 'Ma Perkins' (bred by Jackson & Perkins in 1952). I chose this rose because, although it was no great beauty, it had very cup-shaped flowers like a Bourbon Rose, a form that I very much wanted to obtain in English Roses. For a very rich crimson I chose 'Château de Clos-Vougeot' (bred by Pernet-Ducher in 1920) which, though weak in growth, had an unrivalled reputation in providing roses of this colour.

The Origins and Nature of an English Rose

By the late 1960s and early 1970s, we ourselves were able to introduce a number of roses of the 'English' type, since by that time we were up and running as David Austin Roses. These included 'Wife of Bath', 'Chaucer', 'The Prioress', 'Canterbury' and 'The Knight'—all names taken from Chaucer. They had much of the Old Rose beauty I so much desired, but were unfortunately rather weak in growth and somewhat inclined to disease. It may be that I was too hasty in putting these roses on the market as, at that time, they gave English Roses the reputation of being subject to disease—something that is, today, the exact opposite of the truth. Nonetheless, they proved to be very popular and succeeded in introducing the idea of English Roses to the public.

What I had to do next was to find further parents for a new generation of roses with the vigour and disease-resistance essential to success. At the same time I had to be careful not to lose the particular charm and character of the roses that I already had. With this in mind, I chose the repeat-flowering Old Roses—the Portlands, the Bourbons and the Hybrid Perpetuals. These were often not quite what I was looking for among the Old Roses, but they did have the advantage that they were—at least in some degree—repeat-flowering. When they were crossed with our original roses, the resulting seedlings repeated well and still had the type of blooms I desired. Perhaps the most successful of this group of parents have been the Portland Roses.

In spite of the fact that English Roses were becoming popular and were receiving a great deal of interest from the gardening media, I was still dissatisfied with them. They had much of the beauty and fragrance I desired, but they still did not quite have the vigour and disease-resistance that I would have wished. I was looking for a vigorous, well-rounded shrub that at least had good disease-resistance and eventually, if possible, complete disease-resistance. I therefore turned to the vigorous Rugosa hybrid, 'Conrad

ABOVE 'Wife of Bath'—a cross between the old Hybrid Tea 'Madame Caroline Testout' and a seedling from 'Constance Spry' —was one of our first repeat-flowering English Roses.

Ferdinand Meyer'. For those who know this rose, it might seem an odd choice of parent, since it is not particularly disease-resistant for a Rugosa and is unusual among these roses in having a flower of Hybrid Tea formation. It does have enormous vigour, although it is also rather ungainly in growth. However, I chose this variety more for its parentage than for itself. It is the result of two very different parents, the lovely old Noisette climber 'Gloire de Dijon' with yellow flowers and a vigorous Rugosa hybrid that was almost certainly of pink or red colouring. *Rosa rugosa* is one of the three naturally repeat-flowering species and was likely to be helpful in this respect. It is extremely hardy, very disease-resistant and has an excellent fragrance. My idea was that it might be possible to bring some of the quality of flower of 'Gloire de Dijon' together with the vigour of the Rugosas and combine these characteristics with existing English Roses—while at the same time maintaining and perhaps enhancing their beauty. I cannot pretend that I had very high hopes, but good things often occur in rose breeding by simply casting around in this way.

In the event, I got two very different groups of seedlings, each very useful in their way. It seemed that the 'Gloire de Dijon' half of the ancestry of 'Conrad Ferdinand Meyer' passed on its characteristics separately from those of the Rugosa half—the two seldom came intermingled. So we had two very different types of seedling roses: those that followed the Noisette side of 'Conrad Ferdinand Meyer' and those that followed the Rugosa side of that rose. In fact, we had two new types of English Rose for the price of one. Both proved to be highly beneficial to the family of English Roses as a whole.

The Noisette side gave us the ever-popular 'Graham Thomas', which brought a very rich yellow to the breed, something that we very much needed. It was also typically 'Noisette' in its growth and foliage. The Rugosa side gave us 'Mary Rose', with its true Old Rose flowers and foliage and excellent bushy growth. Both strains were hardy, healthy and vigorous and had unusually good repeat-flowering properties. They subsequently had a very strong influence upon English Roses. 'Mary Rose' is not a typical Rugosa, but it is not difficult to see the influence of that rose at work in it. It has typical Old Rose flowers and suffers little disease. It also

ABOVE 'Conrad Ferdinand Meyer' was important as the parent of the English Roses 'Graham Thomas' and 'Mary Rose'.

OPPOSITE 'Mary Rose' has lovely rose-pink flowers of true Old Rose character. This rose and 'Graham Thomas' were our first widely popular English Roses that led the way towards the English Roses as we know them today.

The Origins and Nature of an English Rose

RIGHT 'Graham Thomas',
an exceptionally rich shade
of yellow and still one of our
most popular English Roses,
performs very well both as
a shrub and as a climber.

has vigorous, branching, twiggy growth and repeat-flowers admirably.

The introduction of 'Graham Thomas' and 'Mary Rose' set the seal upon the popularity of English Roses and opened the way for a whole range of new roses. Since the late 1970s and early 1980s we have introduced a large variety of other parents into the English Roses, resulting in a much greater diversity and many new good qualities. These have included the descendants of *Rosa wichurana*, in the form of the repeat flowering-climbers, which are mainly the descendants of 'New Dawn', the foundation parent in a group called 'Modern Climbers'. These were, themselves, often little more than shrub roses, which made them ideal for the development of a large shrub. Another line of development has been hybrids of the Alba Roses, in an endeavour to obtain a large shrub with something of the special beauty of those roses. Other lines have included certain species in search of greater disease-resistance and other good qualities.

PART

TWO

—

A

GALLERY

OF

ENGLISH

ROSES

T HE PHOTOGRAPHS and descriptions of English Roses represented here are those we regard as the best varieties we have bred to date. Some earlier varieties have been superseded by these—either because of the beauty of their flowers and growth or, more often, for their health and vigour. New varieties are introduced annually. These will, we hope, extend the range and beauty of English Roses in years to come and we expect to include them in future editions of this book.

Alongside each picture is a brief description, in which I endeavour to be as honest as I possibly can. Since these roses are—so to speak—my 'children', you may think that this has not been easy. No one, however, sees the weakness of his roses as clearly as the breeder—and I believe I have given a fairly balanced account. Indeed, I would like to warn the reader against rejecting a rose simply on the grounds that I have noted a fault of one kind or another. All roses—like all people—have their faults. It is simply a matter of weighing up the priorities for your own garden and your own particular taste. Occasionally it may seem that my description does not fit the picture exactly, but it has to be remembered that a rose is not a static thing—it is always changing according to the season; that is to say, early summer or late summer, the prevailing weather conditions, soil, climate and so on.

The perfect rose is an impossibility, if only because tastes often conflict on various characteristics. In fact, the virtues and vices of a rose can also conflict: a thin petal, for example, may provide us with a particularly beautiful rose, but then we have the problem that a very thin petal is subject to damage from damp or the sun. There are many times when we have to strike a balance on certain characteristics.

Fragrance in a rose is a particularly elusive quality: it is made up of a large number of chemicals. Some of these may not be activated under certain

A Gallery of English Roses

conditions: it may be too hot or too cold for the rose to give of its best, or it may be too late in the day and the rose has already spent some of its fragrance. As a result, you may not be able to make a true assessment of a rose's scent on a single acquaintance. Indeed, it is quite possible to get three different scents from three different blooms on the same shrub at the same time.

There is something to be said for a certain amount of inconsistency in a rose. The blooms produced by most varieties range from average to absolute perfection. You never quite know what you are going to get. When you do find that 'perfect' rose, there is an extra pleasure.

It is said that when you choose a rose for your garden, it is best to see it growing first—and this is certainly true. The problem is that you may not be seeing it at its full beauty. The flowers of the most excellent variety may be past their best. You may visit a nursery or garden centre to make a choice and walk past the very rose you would have loved to have, but it happens to be out of flower. There is also something to be said for studying them in a book like this, or a good catalogue, before you choose.

DIMENSIONS As regards growth and the dimensions that we may expect a variety to achieve, I describe English Roses as 'small', 'medium' or 'large', with the occasional 'small to medium' and so on. If a variety is of spreading growth or of taller, narrower growth, this also is noted. Using figures is un-satisfactory since growth varies so much depending on where and how the roses are grown. It can vary even more according to the way in which they have been pruned. It is possible to prune a quite large English Rose hard back and have a comparatively small plant, or light prune it and have a much larger shrub. Also a rose well grown in a warmer climate than ours in the British Isles will nearly always be larger. For all these reasons I have thought it better not to be too precise. Gardeners can adapt what I say as regards dimensions according to their own conditions. By 'small shrub' I mean around about 90cm/3ft in height; by 'medium shrub' 1.2–1.5m/4–5ft; and by 'large shrub' 1.8m/6ft or more. All these figures are open to wide modification according to conditions.

Climbing English Roses require some explanation. It is not always possible to define an English Rose as either a climber or a shrub. Many of them perform equally well in both roles. In the descriptions that follow, the roses in Groups I to IV are Shrub Roses and most are only suitable for growing as shrubs, but a number of them are dual purpose and I include a note describing how they perform as climbers. I categorise all the climbing roses in Group V as English Climbers. They can (with the exception of 'Malvern Hills'), be grown not only as climbers but also as large shrubs in the border.

As regards size of flower, I use a similar system to that which I use for growth. By 'small' flowers I mean about 60mm/2½in across, 'medium' are 60–100mm/2½–4in across and 'large' have a diameter of 100–125mm/4–5in across and sometimes more. Very small, spray-flowered roses like 'Francine Austin' are described as 'pompon' and their flowers are usually 25–40mm/1–1½ inches across. These measurements will vary in very much the same way as do the dimensions of the plant. Here again, cultural conditions, pruning (harder pruning usually producing larger flowers) and climate, will all have their effects.

DIVERSITY & CLASSIFICATION   The Hybrid Teas and Floribundas of the present day have been more and more standardised in the hands of the plant breeder. The rose that used to be a bushy shrub has tended to become a bedding plant. Although flower colours vary widely between one variety and another, in all other respects it is a fairly uniform product. This is unfortunate, particularly since the rose is planted in almost every garden in Britain as well as most other countries throughout the world. We really do not want to meet

A Gallery of English Roses

the same flower again and again as we move from one garden to another. Indeed, this 'sameness' among Hybrid Teas and Floribundas has led to some gardeners leaving all roses out of their gardens.

With the English Roses, we have reversed this process of uniformity by breeding from a great variety of roses both old and new and, more recently, with a number of different species. This has resulted in a collection of roses of widely varying character and beauty, which we believe has added greatly to the pleasure they afford us. The process is ongoing and we expect even greater variation in the years to come. All this diversification has necessitated further classification.

I have divided the English Roses into six groups. *The Old Rose Hybrids* are English Roses that are largely of Old Rose origin. *The Leander Group* includes English Roses that are closer to the Modern Roses in their breeding. *The English Musk Roses* are English Roses and other roses that are partly related to the Musk Roses. *The English Alba Hybrids* are Alba Roses hybridised with English Roses. *The Climbing English Roses* may be related to any one of the above groups. The sixth group comprises *The English Cut-Flower Roses* that have been bred especially for the cut-flower trade.

It is important to emphasise that these groups are by no means clear-cut: one group may overlap with another in some of its varieties. Each group has its own overall character—the 'something' that gives the roses their beauty and is different in each case. I believe that classifying the English Roses in this way enables gardeners to understand their diversity and so enjoy them more completely. It also helps when selecting roses for the garden—one type of rose fits well in one area of the garden, while others are more suitable for another. An introduction to each group describes its overall characteristics and special kind of beauty.

GROUP

I

THE OLD ROSE HYBRIDS

THE OLD ROSE HYBRIDS are the original English Roses, whose evolution I discussed in Chapter 6. The Old Rose Hybrids, it will be remembered, are the result of crossing the early summer-flowering Old Roses with modern Hybrid Teas and Floribundas, with the idea of combining the best characteristics of both: that is to say, the repeat-floweringand wide colour range of Modern Roses with the unique beauty and natural, shrubby growth of the Old Roses. This group is very close to the Old Roses in character although there are varieties that still lean a little towards the modern. All other groups have developed at least in part from this one. More recently we have used the Rugosa Roses as parents in the hope of giving the group even better disease-resistance, while retaining their Old Rose character and developing more bushy, shrubby growth and further enhancing their repeat-flowering qualities.

The colours of the Old Rose hybrids range from white, through blush to pink, deep pink and crimson. At present most of our best crimson English Roses are to be found in this group. We have recently introduced the colour yellow, without losing the essential Old Rose character of the group, and so far we have two varieties, 'Windrush' and 'Jude the Obscure'. Both are, we feel, sufficiently 'Old Rose' to be included here.

The fragrance of the Old Rose hybrids is, as might be expected, predominantly that of the Old Roses, although this is often mixed with tea scent, myrrh, lily of the valley, lilac, almond blossom and others. The fragrance is nearly always very strong.

The flowers of these roses are not flamboyant but have the unassuming charm of the Old Roses. They usually have a rosette shape with numerous delicate, rather translucent petals, often closely packed to give them a pleasing, glowing effect. Like those of the Old Roses, the flowers have a certain softness and a neatness of character that is most appealing. They are, in fact, well removed from Hybrid Teas and Floribundas in appearance.

The foliage of the roses in this group tends to be more opaque in character, similar to the Old Roses, although it frequently leans towards the Modern. Where there is an element of *Rosa rugosa* in their make-up, the leaves are often more elongated and the leaflets more widely separated and, somewhat surprisingly, more 'Old Rose' in appearance.

OPPOSITE 'Eglantyne', a typical variety of the Old Rose Hybrids, harmonising nicely with *Scabiosa columbaria* subsp. *ochroleuca* and *Salvia nemorosa* 'Ostfriesland'.

A Gallery of English Roses

Barbara Austin

Along with 'Gertrude Jekyll' and 'The Countryman', this rose is closely related to the old Portland Roses, which were the first repeat-flowerers to be introduced after the China Roses. The Portland Roses are, to my mind, among the most beautiful of the repeat-flowering Old Roses and also some of the most fragrant.

'Barbara Austin' is very close to a Portland Rose in overall character. It is very strong—forming a thicket of growth—and has typical Old Rose foliage and flowers, with many petals opening wide to reveal a rather informal bloom of medium to large size. The colour is a delicate blush pink. The fragrance is particularly fine—a mixture of Old Rose and lilac. The rose's only weakness is a tendency to send up the occasional strong shoot that fails to flower, rather like a non-repeating Old Rose. Cut back by about two thirds as soon as they develop, these will become strong flower shoots later in the season. 'Barbara Austin' will form a medium-sized shrub.

Named after my sister, Barbara Stockitt, who, with her husband Philip, is proprietor of Barbara Austin Perennials Ltd at West Kington in Wiltshire, where they have a beautiful garden.

Variety AUSTOP | US Patent no. 11423 | Introduced 1997

Brother Cadfael

'Brother Cadfael' bears some of the largest blooms to be found among the Old Rose Hybrids. They are of a deeply cupped, chalice shape, the petals being somewhat incurved. The individual flowers are held upright on a bushy plant. Later in the season they may be shallower in shape and perhaps a little smaller, but are equally beautiful. Their colour is a soft rose pink. The growth, which is of the Old Rose type, is particularly strong, forming a fine, medium-sized shrub. They have a rich old Rose fragrance.

Named after the hero of Ellis Peters' detective stories, which are set in medieval Shropshire.

Variety AUSGLOBE | US Patent no. 8681 | Introduced 1990

Charles Rennie Mackintosh

This rose has flowers of unique colouring for an English Rose, a lovely shade of soft lilac that sometimes varies to a lilac pink according to weather conditions—shades we would like to develop further among our roses, particularly because they are so useful for mixing with other colours in the rose border and in flower arrangements. The flowers are quite large, cupped in shape and well filled with petals; they have a light Old Rose fragrance, fittingly with aspects of lilac and almond blossom. The growth is upright, of medium size, tough and thorny, with typical Old Rose foliage. It is exceptionally free-flowering—and we get to like this rose more and more as the years go by.

Named after the architect, painter and designer, in conjunction with the Charles Rennie Mackintosh Society and the City of Glasgow Parks and Recreation Department. Mackintosh (1868–1928) frequently used stylised roses in his designs which were often similar to this variety.

Variety AUSREN | US Patent no. 8155 | Introduced 1988

Cottage Rose

A bushy, upright shrub of short growth that flowers with remarkable freedom and regularity, more so than most English Roses. The individual flowers are typically Old Rose, medium in size, rosette shaped and slightly cupped, their colour being a lovely warm pink. They have a delicate Old Rose fragrance with a suggestion of almond and lilac, which is surprisingly diffusive in a warm room. Very occasionally and rather unexpectedly, it sends up a long climbing shoot—this should be removed.

Variety AUSTGLISTEN | US Patent no. 8671 | Introduced 1991

Eglantyne

I have long regarded this rose as one of the most charming of the English Roses, at least in the individual flower. It is closely related to 'Mary Rose', to which it is similar, but has even greater refinement. The growth, too, is similar but more upright and not quite so vigorous. Its flowers are of a lovely soft pink, paling towards the edges and of the most perfect rosette formation; they have much of the kind of beauty we usually associate with the old Alba Roses—always one of the highest compliments I can give any rose. It is sweetly scented: a charming and delicate Old Rose fragrance. 'Eglantyne' forms a short to medium sized shrub.

Named after Eglantyne Jebb, a Shropshire lady who founded the Save the Children Fund after the First World War.

Variety AUSMAK | US Patent no. 9526 | Introduced 1994

Falstaff

*We are always looking to breed good crimson English Roses, but these
are not easy to achieve. 'Falstaff' is one of the finest we have bred to date.
It bears quite large flowers of shallow, cupped shape. They are of exquisite
form and quality with numerous petals which interfold at the centre to produce a lovely, glowing effect within an enclosed saucer of outer petals. The
colour is deep crimson at first, paling towards the outer petals, the whole
flower slowly turning to a pleasing shade of magenta-crimson. Unlike so
many crimson roses, it is of strong growth, bushy and rather upright, with
the flowers nodding nicely on the stem. The foliage is quite large and rather
more 'modern' than Old Rose. It has a wonderful rich Old Rose scent.*

Named after the character in Shakespeare's Henry IV.

AS A CLIMBER

*'Falstaff' is not a natural climber, being rather too bushy, but we have few
red English Climbers and it will grow to medium height when planted
against a wall.*

Variety AUSVERSE | US Patent no. 13315 | Introduced 1999

Gentle Hermione

This is a rose of mixed parentage, having strains of both English Musk and Leander in its make-up, as well as that of the Old Rose hybrids. However, its appearance leans towards the Old Roses. The growth is of about medium size, becoming quite broad with time. The leaves are tinged red at first, becoming green and then dark green with age. The flowers, as the photograph shows, are soft pink, the colour softening towards the edges with age. They gradually open to form shallow cups. They are a perfect shape at all stages, the whole effect being totally charming. They have a strong Old Rose fragrance with hints of myrrh. The disease-resistance is particularly good.

Hermione was the faithful wife of Leontes, King of Sicilia and mother of Perdita in Shakespeare's 'The Winter's Tale'.

Variety AUSRUMBA | Introduced 2005

Gertrude Jekyll

'Gertrude Jekyll' is closely related to 'The Countryman', both being the result of a Portland Rose cross. By sheer chance they appeared alongside each other in our Trial Grounds. 'The Countryman', however, is a spreading rose, while 'Gertrude Jekyll' is more upright in growth. It has large, deep pink flowers with very much of an Old Rose character. Its flowers are not the most perfectly formed, but this does not seem to rob them of their beauty. Both the foliage and growth are close to that of a Portland Rose, with typical Portland leaves and widely spaced leaflets of a light green colour. It is very robust and free-flowering and has a very strong, rich Old Rose fragrance of a high quality. It forms a medium-sized shrub.

Named after the English garden designer and author, Gertrude Jekyll (1843–1932).

AS A CLIMBER

Rather surprisingly, this rose has proved to form quite a good climber.

Variety AUSBORD | US Patent no. 7220 | Introduced 1986

Harlow Carr

'Harlow Carr' was introduced in the same season as 'Rosemoor' and both bring something new to English Roses. Rather like the Old Roses Rosa centifolia 'De Meaux' and R. 'Petite de Hollande', 'Harlow Carr' has small, almost miniature flowers of perfect Old Rose formation; starting as delightful little cups they develop into a cupped rosette-shaped flower, with a button eye at the centre. The colour of 'Harlow Carr' is a lovely, even, medium rose pink and it has a strong Old Rose fragrance that has been described as reminiscent of rose-based cosmetics. The growth is exceptionally vigorous and healthy, developing into a shapely, well-rounded, spreading shrub, bearing its flowers almost to ground level. The foliage is of the Old Rose type, bronze at first, later becoming green. All in all, 'Harlow Carr' is almost the ideal English Rose of gem-like beauty. I hope that gardeners will not pass it over simply because the flowers are small. It will form a medium-sized shrub.

Named after Harlow Carr in Yorkshire, the most northerly of the Royal Horticultural Society's gardens, to mark the two hundredth anniversary of the formation of the Society.

Variety AUSHOUSE | Introduced 2004

Heather Austin

This rose forms a vigorous, medium to tall shrub of rather upright growth. Its flowers are unusual among the Old Rose hybrids in that they are of cupped, chalice-like shape, incurving a little and often showing their golden stamens within. They are truly Old Rose in character and of a deep rose pink and held on long stems, making them an ideal addition to a bowl of English Roses. The leaves are large, light green and slightly glossy. There is a delicious, strong Old Rose fragrance.

Named after my sister, Heather Austin, now Heather Coulter.

Variety AUSCOOK | US Patent no. 10618 | Introduced 1996

Hyde Hall

This variety stands alone among English Roses; however, it seems to fit in best among the Old Rose Hybrids. Perhaps its most significant characteristic is the fact that it forms a very large shrub and manages to combine this with an exceptional ability to repeat-flower. Very few repeat-flowering roses of any kind achieve this. We cannot usually have generous size and frequent repeat-flowering. It seems that no sooner has it produced one lot of flowers than it is sending shoots out in all directions. The flowers are of medium size and not particularly glamorous, but they do have a simple beauty. This rose's excellence lies more in its effect as a shrub than in the individual flowers. Medium pink and of rosette formation, these are produced in small and large sprays. Although the fragrance is relatively light, it is nonetheless delightfully warm and fruity. The foliage is fine and pointed, and rather similar to Rosa canina, *the dog rose. It is very healthy. 'Hyde Hall' forms a fine specimen shrub.*

Named after the Royal Horticultural Society's garden in Essex, which is the Society's main rose garden and includes many English Roses.

Variety AUSBOSKY | US plant patent applied for | Introduced 2004

John Clare

'John Clare' is an exceptionally fine English Rose which flowers with out-standing continuity throughout the summer. It is almost as good in the autumn as it is in early summer. Its only weakness is that it has no more than a very slight fragrance—at least in the British climate. The flowers are cupped in shape, full-petalled and of a deep, glowing pink. It forms a low, arching, nicely rounded shrub. The foliage is unusually glossy for a rose of this group. 'John Clare' is an altogether reliable rose that is an excellent garden plant.

Named after the poet John Clare, who was a farm worker but eventually became one of the foremost nature poets in the English language.

Variety AUSCENT | Introduced 1994

Jude the Obscure

We have, among the Old Rose Hybrids, all the colours from white through to pink and on to crimson, mauve and purple—but we had no yellow flowers until we introduced the variety 'Windrush', from which we later bred 'Jude the Obscure'. Both these varieties are of a pleasing shade of soft yellow. Since there were no yellow roses among our Old Rose Hybrids, we had to go outside the group to find breeding material for this colour.

'Jude the Obscure' is, we think, of sufficiently Old Rose character to be included in this section. The flowers are imposing: large, deeply cupped and incurved, with numerous small petals within. Their colour is of a rather stronger yellow inside than outside the flower. They have a strong and unusual fragrance with a fruity note, reminiscent of guava and sweet wine, which we find particularly pleasing. The growth is short, strong, upright and bushy, with light green leaves—all in all, quite typical of the English Old Rose hybrids.

Named after Thomas Hardy's novel.

Variety AUSJO | US Patent no. 10757 | Introduced 1995

Kathryn Morley

A lovely, deeply cupped rose of true Old Rose character. The flowers are quite large with many soft pink inner petals, the outer petals eventually turning back and becoming almost white—creating a beautiful effect. They have a pleasing fragrance: a mixture of Old Rose and myrrh. Later flowers may be more shallowly cupped. The foliage leans a little towards that of the Old Rose. 'Kathryn Morley' forms a small to medium-sized shrub of average vigour. The photograph shows the blooms at an open stage—earlier they are much more cupped.

Mr and Mrs Eric Morley of the Variety Club of Great Britain bought the right to name this rose at a charity auction and named it after their daughter, who died at the age of eighteen.

Variety AUSVARIETY | US Patent no. 8814 | Introduced 1990

L. D. Braithwaite

There are several dark red English Roses, but the flowers of this variety are a lighter, brighter crimson. They are loosely formed and usually open wide and flat. 'L. D. Braithwaite' is related to 'Mary Rose' and has many of the excellent qualities of that variety. It flowers freely and with remarkable continuity, particularly for a red rose. Quite the opposite of most roses, it has little scent until the flower ages, when it develops a charming Old Rose fragrance. It is one of the best red varieties for the mixed border, forming a medium-sized bushy, twiggy shrub that fits in easily with other plants.

Named after my father-in-law, Leonard Braithwaite.

Variety AUSCRIM | US Patent no. 8154 | Introduced 1988

Mary Magdalene

This is a modest rose with a gentle charm that is very pleasing, the whole effect being one of simplicity and softness. The flowers are slightly less than medium sized, a soft apricot pink with silky petals arranged around a button eye. It forms a small, compact shrub with low, arching branches and typical Old Rose foliage. It has a particularly beautiful Tea Rose fragrance with just a hint of myrrh. A lovely rose that benefits from a little extra feeding and care.

Named after our local church of St Mary Magdalene, Albrighton.

Variety AUSJOLLY | Introduced 1998

A Gallery of English Roses

Mary Rose

Not only was 'Mary Rose'—along with 'Graham Thomas'—one of the first English Roses to become widely popular after the introduction of 'Constance Spry', but it has also played an important part in the development of the English Roses as a whole. It forms a medium-sized, twiggy shrub and produces flowers with more than usual continuity. These are of typical Old Rose rosette formation and of medium size with a strong rose-pink colour. They have a light Old Rose fragrance with a hint of honey and almond. In warm seasons and in warmer climates than that of the UK, the fragrance can be much stronger. 'Mary Rose' is still one of our better roses, after more than two decades of further breeding.

'Mary Rose' has an element of Rosa rugosa in its parentage and a tendency to sport—something we often find in more extreme hybrids of a new strain. At times it will produce a sport that is half pink and half white. The unwanted growth can easily be cut out. It has produced two very good roses in this way, 'Winchester Cathedral' and 'Redouté'. Rosa rugosa has also brought good disease-resistance.

Named on behalf of the Mary Rose Trust to mark the recovery in 1983 of Henry VIII's flagship from the Solent after more than four hundred years. Treasures from this ship are still being recovered from the sea.

Variety AUSMARY | Introduced 1983

Winchester Cathedral

This is a white sport of 'Mary Rose'— which was very welcome, as we are rather short of white English Roses. It has all the virtues of 'Mary Rose' and is the same in every way except colour. It is probably our best white to date.

Named in aid of the Winchester Cathedral Trust.

AUSCAT | US Patent no. 8141 | 1988

Redouté

As with 'Winchester Cathedral', 'Redouté' is exactly the same as 'Mary Rose', except that its colour is a lovely soft shade of pink. It is, perhaps, the most beautiful of the three roses.

It seemed only right that there should be a rose bearing the name of Pierre-Joseph Redouté (1759–1840), one of the first in a long tradition of rose painters in France and certainly the most accomplished.

AUSPALE | US Patent no. 8789 | 1992

A Gallery of English Roses

Miss Alice

'Miss Alice' bears flowers that are soft, warm pink at first, the outer petals soon turning to a pale pink which gradually spreads over the whole flower as it ages. They are about 63 mm / 2½ in a cross, of neat rosette formation and well filled with small, closely packed petals. It has a lovely, well-rounded Old Rose fragrance with additional hints of lily of the valley. It makes a small, neat shrub that is particularly good for a rose bed or for the front of a border.

Named after Miss Alice de Rothschild, who created a beautiful rose garden at Waddesdon Manor in Buckinghamshire, now owned by the National Trust. The rose garden was reinstated in 2000 with roses supplied by our Nurseries.

Variety AUSJAKE | Introduced 2000

A Gallery of English Roses

Noble Antony

This rose bears full flowers of an attractive shade of deep magenta-crimson whose outer petals gradually turn back to form a deeply domed shape. The flowers are set against dark green foliage, creating an overall dark effect that is most pleasing. The rose's beauty is best appreciated when it is cut for arrangement in the house, where it has excellent lasting qualities. It has the lovely Old Rose fragrance that we expect of a rose of this colour and was awarded the prize for fragrance at the Glasgow Trials. It forms a small shrub, suitable for bedding or for the front of a border.

Named after Mark Antony in Shakespeare's Julius Caesar.

Variety AUSWAY | US Patent no. 10779 | Introduced 1995

Rosemoor

Much of the description of 'Harlow Carr' (page 88) applies equally to this rose. The flowers of 'Rosemoor', however, are a little smaller, no more than 50mm / 2in across. Each bloom is like a perfect miniature Old Rose. Their colour is a soft pink—deeper at first, gradually becoming paler. They are held in small and large sprays but the rose does not look like a spray-flowered variety as the blooms tend to open in succession and each flower is seen in all its beauty. It has a lovely fragrance which is described by our expert as 'a rose scent with hints of apple, cucumber and violet leaf'. Not only is the individual flower charming, but the overall effect of the shrub is also very beautiful. It forms a small to medium-sized, perfectly rounded shrub covered with small flowers almost to the ground, giving the effect of a Japanese cherry in full bloom. To complete its list of virtues, it is in our experience almost completely free of disease.

Named after the Royal Horticultural Society's Devon garden.

Variety AUSTOUGH | Introduced 2004

Sharifa Asma

An English Rose of great beauty. Its blooms are rather more than medium-sized, prettily cupped at first, opening into a rosette shape with numerous folded petals within. The colour is a lovely soft pink mixed with palest blush on the underside of the petals, creating a most charming effect. The growth is bushy and rather upright; the foliage is of the typical Old Rose type, perhaps due to a small percentage of Rosa rugosa *in its parentage. The fragrance can be almost overpowering, with hints of white grape and mulberry. This has understandably proved to be one of the most popular English Roses. Its disease-resistance is satisfactory, though not so good as that of some of our more recent varieties.*

Named after an Omani princess at the request of her family.

Variety AUSREEF | US Patent no. 8143 | Introduced 1989

Sophy's Rose

'Sophy's Rose' has flowers that are a little different from other English Roses. The petals are small at the centre and increase in size, by degrees, until the outer petals are quite large, resulting in an unusually flat bloom, almost like a zinnia. Their colour is light crimson. The general impression is of an airy lightness. This is a very productive rose of short, twiggy growth, making it suitable both for rose beds and for the front of the border. It has a light Tea Rose fragrance. It is very healthy and repeat-flowers extremely well.

Named after Sophy, daughter of Wendy Fisher, who founded the Dyslexia Institute in 1972.

Variety AUSLOT | US Patent no. 11422 | Introduced 1997

Tess of the d'Urbervilles

This rose bears large flowers of a lovely cerise crimson. They are beautifully formed with loosely packed petals intertwined at the centre; the outer petals later turn back to form a dome-shaped flower. The blooms are produced early with great freedom, often bowing down the branch with their weight. Later flowers are less heavy, but have their own particular beauty. There is a pleasing Old Rose fragrance. The growth is strong and healthy with large, dark green leaves. It forms a medium-sized shrub.

    Named after Thomas Hardy's novel.

AS A CLIMBER
This variety forms a reliable climbing rose of 1.8–2.4m / 6–8ft.
Variety AUSMOVE | Introduced 1998

The Countryman

A unique rose that is, like 'Gertrude Jekyll', the result of crossing one of our earlier English Roses with a Portland Rose. It is upright at first, but the branches soon fall to provide nicely spreading growth which builds into a shapely mound. Its plentiful pale green leaves come right up to the flowers, displaying them to perfection—the individual leaf having widely spaced leaflets, revealing the Damask ancestry of the Portland Roses. Its growth is short and rather wider than it is tall. The flowers are well formed, with numerous narrow petals, opening to a shapely, flat rosette. Their colour is a lovely, even rose pink. They have a delicious Old Rose fragrance, sometimes with overtones of strawberry. In every way this is an excellent, healthy and very reliable rose—and one of my personal favourites.

Named on behalf of The Countryman *magazine.*

Variety AUSMAN | US Patent no. 7556 | Introduced 1987

The Dark Lady

A highly individual English Rose, in part due to the fact that it has some Rosa rugosa in its ancestry. Its branches do not so much arch as grow outwards, sometimes almost at a right angle. This, rather surprisingly, gives us a good, shapely shrub which is small at first, but gradually grows quite big. It has large, rough-textured, dark green leaves, bearing some resemblance to those of its Rugosa parentage. Its flowers, which hang elegantly on the stem, are of a dark, dusky crimson and open flat to reveal a large, quite loosely petalled flower with something of the character of a tree peony. There is a particularly rich Old Rose fragrance. Unfortunately, in Britain this rose has a tendency to blackspot and requires spraying. However, in warmer climates with less moisture in the air, it is remarkably free of all diseases. Indeed in the southern USA it is one of the most successful English Roses.

The name is taken from the mysterious 'dark lady' of Shakespeare's sonnets.

Variety AUSBLOOM | US Patent no. 8677 | Introduced 1991

The Mayflower

'The Mayflower' is one of the most important English Roses we have bred since we introduced our first repeat-flowering varieties in the early 1970s. It not only has the character of flower and growth we look for in English Roses—and this in a very high degree—but it is, as far as we can tell, almost completely free of disease. This is something that can be said of very few roses of any kind—particularly repeat-flowering roses.

'The Mayflower' is not a flamboyant rose and you might easily miss it when walking around a garden for the first time, but it has a simple charm that is most pleasing. The flowers are of small to medium size and of slightly domed, rosette formation, the petals turning back at the edges. They have a beautiful and very strong Old Rose fragrance. The growth is bushy and freely branching, producing its flowers in a sprightly manner, well above the foliage, with exceptional regularity throughout the summer. Starting very early—often in late May in Britain, long before other English Roses are in flower—it is seldom without flowers until the onset of winter. It has dainty, almost species-like leaves which are light green tinged with copper at first, later becoming dark matt green. It forms a small shrub initially, but gradually builds up into one of medium size. If it has a fault, it is that some flowers may fail to develop to the full—perhaps because of the great effort its puts in over the season. There may also be a tendency to red spider in warmer climates than that of Britain.

Named after the ship that carried the first English settlers to America, to mark the launch of our branch nurseries at Tyler in Texas.

Variety AUSTILLY | US plant patent applied for | Introduced 2001

William Shakespeare 2000

There can be little doubt that this is the most beautiful, deep crimson English Rose we have so far developed. Its flowers are the very essence of what we feel a deep crimson English Rose should be. They are large and open flat, the petals turning back a little at the edges to form a perfectly formed rosette shape. There are very many intertwining petals which start a pure, deep crimson, then gradually take on many different shades of purple, violet and mauve, creating the most beautiful, ever-changing effects. The petals have a soft, velvety texture which is most appealing. They are displayed against dark green foliage, with which they harmonise ideally. The growth is broad and quite short.

This rose seems to require quite a lot of pruning—the removal of weak and old growth, together with some shortening of the longer branches. It does better in some seasons than others, but is always very worthwhile.

Named after the playwright who was voted 'British Man of the Millennium' by listeners to BBC Radio 4's Today *programme.*

Variety AUSROMEO | US Patent no. 13993 | Introduced 2000

A Gallery of English Roses

Corvedale

A tall, rounded shrub, sending up long branches, making it ideal for the mixed border. Its flowers are of a clear rose pink of medium size and open, cupped formation, displaying their prominent stamens to excellent effect. This is a form of flower that we are anxious to develop among English Roses. It has that strong myrrh fragrance often found in English Roses. This is a reliable, trouble-free rose that will form a large shrub.

Named after a most beautiful valley running parallel to Wenlock Edge in the depths of the Shropshire countryside.

AUSNETTING │ 2001

Lochinvar

This is not a true member of the English Old Rose Group. It is a hybrid of the Scottish rose (Rosa pimpinellifolia) and retains its characteristic hardiness, dense growth, small leaves and charming little flowers —but has the additional advantage of continuing to flower throughout the summer. Its flowers are a fresh, soft pink and have a light fragrance.

Named after the poem by Sir Walter Scott.

AUSBILDA │ 2002

Glamis Castle

'Glamis Castle' has pure white, cup-shaped flowers of true Old Rose character. Its growth is short and bushy and it bears flowers with exceptional freedom and continuity. They have a strong myrrh fragrance. Unfortunately, it is not very resistant to blackspot, but with suitable spraying it is quite satisfactory and is particularly useful as there are few white English Roses.

Named after the legendary setting of Shakespeare's Macbeth *and the birthplace of the late Queen Elizabeth the Queen Mother.*

AUSLEVEL │ US Patent no. 8765 │ 1992

Mistress Quickly

This is not a glamorous rose, but it does have a simple beauty and is very useful in the garden. It forms a tall, remarkably tough, bushy shrub and bears small, semi-double, lilac pink, lightly fragrant flowers well above its foliage. It is, as far as we know, entirely free from disease. Ideal for the back of the mixed border, this rose is also useful where it has to cope with more than usual competition from other plants or shrubs.

Named after the proprietor of the Boar's Head Tavern in Shakespeare's Henry V.

AUSKY │ US Patent no. 10617 │ 1995

Mrs Doreen Pike

This is a modest rose, but one of considerable garden value. It is extremely hardy, will grow under the most difficult conditions and is completely resistant to disease. It forms a neatly domed shrub bearing full, pink, rosette-shaped flowers of medium size that give the impression of mingling with the small leaves. There is a lovely, strong Old Rose fragrance.

Named after Doreen, who managed our offices for many years, in recognition of her contribution to the success of the Nursery.

AUSDOR | 1993

Prospero

The flowers of 'Prospero' are very beautiful and of a rich, deep crimson which fades to an equally rich mixture of purple and mauve. Indeed, no other crimson English Rose has quite such perfectly formed flowers, except 'William Shakespeare 2000'. Unfortunately, 'Prospero' is a very weak grower, and is worth growing only if you are prepared to give it generous treatment, although it is probably stronger in climates warmer than the British Isles. It has a deep, fruity Old Rose fragrance.

Named after the duke in Shakespeare's The Tempest.

AUSPERO | 1982

Portmeirion

'Portmeirion' bears medium-sized, perfectly formed flowers of rich glowing pink. The petals recurve slightly to form a charming, nicely domed flower with a rich Old Rose fragrance. It gradually builds up into a low-growing, mounded shrub. There is some tendency to blackspot, but this is easily controlled.

Named for Susan Williams-Ellis and Euan Cooper-Willis, founders of the Portmeirion Pottery. Susan designed a beautiful range of pottery using English Roses as motifs.

AUSGARD | 1999

The Herbalist

This variety is not unlike the famous old Gallica Rose 'Officinalis' or Apothecary's Rose (now Rosa gallica var. officinalis). Although it is not quite as beautiful, it has the advantage of flowering repeatedly throughout the summer. The flowers are semi-double and deep rose pink with a good boss of stamens. It is a good, reliable garden rose of low, bushy growth that fits in easily with other plants. There is only a light fragrance.

AUSSEMI 1991

Trevor Griffiths

This is a beautiful rose of true Old Rose character. It is rather in the Gallica Rose tradition with flat, rosette-shaped flowers and perfectly arranged petals that are deep pink at the centre and paler at the edges, giving a two-tone effect. There is a strong and very beautiful rose fragrance reminiscent of a fine, old claret. 'Trevor Griffiths' has an excellent, low, bushy habit, holding its flowers high on the bush in a lively manner. The foliage is of a pleasing, dusky green and of a rough texture, marrying well with the darkness of the flowers. It forms a medium-sized shrub. It has good disease-resistance. I like this rose better every year.

Named after my good friend Trevor Griffiths, the New Zealand rose grower, who played a large part in the reintroduction of Old Roses to that country.

AUSOLD | 1994

Windrush

The first yellow rose we bred in this group, 'Windrush' is the result of a cross between an Old Rose hybrid and a species hybrid. It has quite large, semi-double, wide open flowers of soft yellow colouring and a large boss of yellow stamens. It forms a medium-sized shrub with broad growth and soft green foliage. Later on it tends to produce hips at the expense of flowers and you might decide to remove these to encourage further flowering. There is a light, spicy Musk fragrance.

Named after the river in southern England.

AUSRUSH | 1984

Wenlock

This is a strong, reliable rose of vigorous growth—qualities that are not always found among red roses. The flowers are large, mid crimson in colour and rather cupped in shape. I would not place 'Wenlock' among the most beautiful of the English Roses, but it is worth including for its reliability. It forms a medium shrub with plentiful, dark green, disease-resistant foliage. There is a rich Old Rose fragrance.

Named after Much Wenlock, a beautiful mediaeval town near our Nurseries.

AUSWEN | 1984

Wild Edric   New '05 97

This rose might be mistaken for a Rugosa —and, indeed it is in part related, but there are important differences. The foliage is dark green on top and paler green underneath, clean cut, more pointed in shape and somewhat smoother in texture than that of a Rugosa Rose. The thorns are long, thin and pale green. It forms a tall, bushy, upright shrub. The flowers are large, semi-double, substantial and held in small, close clusters, opening in succession. Their colour is a deep pink with a purple tinge at first, becoming a pale lilac-pink, with a contrasting bunch of golden stamens. They have a light fragrance. This is a very tough and entirely reliable rose that will grow well even in poor conditions. It is seldom without flowers.

Named after a Saxon nobleman.

AUSHEDGE | 2004

A Gallery of English Roses

GROUP II

THE LEANDER GROUP

As we crossed Old Roses with Modern Roses, it was inevitable that some of the resulting seedlings would lean towards the former in overall character, while others would lean more towards the latter. It seemed to us that it might be worthwhile developing a group of roses that, while their flowers were of the Old Rose type, their foliage and growth had more of the character of a Modern Rose—rather as one finds in the old Bourbon Roses. For this purpose we turned to the so-called Modern Climbers which, as I explained in Part I, are related to the Wichurana ramblers by way of 'New Dawn', a repeat-flowering sport of the very strong and healthy 'Doctor W. Van Fleet'. Almost all the Modern Climbing Roses are very resistant to disease and all repeat-flower well. Most of them are not very tall and some are, in fact, little more than large shrubs. All these factors made them ideal for the development of our new strain of English Roses.

We decided on the variety 'Aloha', bred by Boerner in 1949, as our foundation parent. This excellent variety is really much nearer to a shrub than to a climber; indeed, we have often grown it as such. It is very disease-resistant and repeat-flowers admirably. Also very important, from our point of view, is the fact that it has nicely cupped flowers of Old Rose appearance and a particularly strong fragrance. It has lovely, clear pink colouring. For all these reasons, it was ideal for our purpose.

'English Wichurana Hybrids' sounded a bit too much of a mouthful, so we decided to gather our new roses under the title the Leander Group, after one of the earliest introductions. Roses in this group are nearly all particularly strong and healthy in growth. They usually form larger shrubs than do the Old Rose Hybrids and have long, often arching growth which gives them a certain grace. Their flowers vary from rosette shape to a deeply cupped formation and are held gracefully, often nodding a little upon the branch. Their fragrance is nearly always strong and widely varying—Old Rose, tea, myrrh, musk and so on—often with an added fruity quality.

All in all, the Leander Group forms a very useful range of roses that does well wherever roses will grow at all. If they lose a little to the other groups in Old Rose charm, they gain much in sheer flowering ability and showiness. One of their chief beauties is the elegant grace of their growth. Both flower and leaf have an added glistening quality which is most pleasing. Of all the English Roses, they are the first to catch the eye.

OPPOSITE 'Golden Celebration', an excellent example of a rose of the Leander Group, manages to be large-flowered and showy without being in any way ungainly.

A Gallery of English Roses

Abraham Darby

One of the largest-flowered of all the English Roses and the only one to have nothing of the Old Rose in its parentage—except of course, like all roses, many generations back—'Abraham Darby' nonetheless has all the characteristics of a typical Leander rose. The flowers are cup-shaped and their colour is complex: a mixture of yellow and pink, with the pink dominating, the outer petals being tinged with yellow towards the edges. This is a vigorous, medium-sized shrub of arching habit, with large, polished, rather modern leaves; flower, growth and leaf are all in proportion and never clumsy. There is a strong, fruity, rose fragrance with a raspberry sharpness.

There is no particular virtue in size for its own sake, but we need large-flowered varieties in English Roses, as they do bring variety and a different kind of beauty. Large-flowered English Roses can fulfil a similar role in the garden to that of the tree peony and, at the same time, flower throughout the summer. This rose has played an important part in the development of the Leander Group.

Named after Abraham Darby (1678–1717), one of the founders of the Industrial Revolution, which began in Shropshire.

Variety AUSCOT | US Patent no. 7215 | Introduced 2000

Alan Titchmarsh

This variety bears quite large, deeply cupped, full-petalled, slightly incurved flowers that are well filled with petals. The outer petals are pale pink, the centre petals of a much stronger, glowing warm pink. The side flowers nod nicely on the branch and are held in small groups of three or four, while the main branches from the base may consist of much larger sprays with the flowers well apart, making an almost candelabra effect. They have a warm Old Rose fragrance with a hint of citrus. 'Alan Titchmarsh' forms a quite large shrub of elegant, arching growth, providing a most satisfactory overall picture. The leaves are red at the bud stage, soon turning to glossy, dark green with seven to nine leaflets. This rose has very good disease-resistance.

Named after the broadcater and writer, who has given so much pleasure and help to gardeners over the years.

Variety AUSJIVE | Introduced 2005

Benjamin Britten

A little different in character from the other roses of this group, this is a bushy shrub of somewhat random, intertwining growth, the stems being thin with rather small, light green leaves. The flowers are light in weight, fully double and of no more than medium size, usually with a button eye. The colour is a mixture of orange and red: a beautiful shade, particularly when the flowers are cut for arrangement in the house. Otherwise 'Benjamin Britten' is a typical rose of the Leander Group—medium in height, bushy and slightly arching, with small, pointed foliage. It is not, perhaps, so free-flowering as one would wish, but unusual and interesting, particularly for its colour.

Named after the British composer.

Variety AUSENCART | US plant patent applied for | Introduced 2001

Crown Princess Margareta

This is a large, vigorous and well-rounded shrub with slightly arching growth. The flowers are a rich apricot-orange. They are a little over medium size and a neat, shallow cup shape. They are beautifully formed, the petals mingling to excellent effect at the centre and the outer petals falling back and becoming paler. The flowers have a strong, fruity fragrance of the Tea Rose type. This is a tough, reliable rose with excellent growth and plentiful glossy foliage. It provides a lovely splash of colour in the garden.

Named after Crown Princess Margareta of Sweden, the granddaughter of Queen Victoria, an accomplished landscape gardener, who created the garden at the Summer Palace of Sofiero in Helsingborg, Sweden.

AS A CLIMBER

'Crown Princess Margareta' forms an excellent climber when grown on a wall.

Variety AUSWINTER | US Patent no. 13484 | Introduced in 1999

Geoff Hamilton

This rose forms a large shrub that is vigorous and healthy in growth. It produces plentiful blooms of a rather unusual formation—they are chalice-shaped rather than cupped and the centre of the flower is filled with petals to provide a most pleasing effect. The photograph is of a bloom that is not fully out. The inside of the flower is a soft, warm pink; the outside is pale pink. The foliage is closer to that of a Hybrid Tea than of an Old Rose. The scent is Old Rose with a hint of apple. 'Geoff Hamilton' is an excellent rose for the back of a border, where it will compete well with other plants.

Named after the much-loved television gardener, who died in 1996.

Variety AUSHAM | US Patent no. 11421 | Introduced in 1997

A Gallery of English Roses

Golden Celebration

This is, perhaps, the most magnificent of the English Roses, with very large, golden-yellow, deeply cupped flowers beautifully poised on gracefully arching stems. In spite of their great size, the flowers are never ungainly. In fact, this rose provides a fine illustration of how even the largest blooms do not have to be clumsy. Everything depends on how they are attached to the shrub. Later in the season, the flowers are not quite so big and are often closer to a cupped rosette shape. The foliage, which also is large, is glossy and of a pale green, a colour that seems ideally to suit that of the flowers. 'Golden Celebration' forms a quite large shrub which complements the flowers perfectly, everything being in proportion. Its health is excellent. The fragrance is strong—tea-scented at first, later developing hints of Sauternes wine and strawberry.

Variety AUSGOLD | US Patent no. 8688 | Introduced 1992

Grace

This is one of the most beautiful of the Leander Group. The growth is broad and arching, building up to form a truly graceful, mound-like shrub. Its flowers are quite large and perfectly formed, held on the end of long, sweeping branches. They begin as a cup shape, quickly developing into dome-shaped flowers consisting of narrow, pointed, slightly fluted petals—making it somewhat different from other roses. The colour is a lovely mixture of shades of apricot, deepening towards the centre. There is a delicious, warm, sensuous fragrance. It has quite thin flower stems, tinted red at first, becoming green, and gracefully hanging, rather pointed leaves. The growth is healthy and vigorous, forming a medium-sized shrub, and it repeat-flowers extremely well.

'Grace' is in every way a lovely rose. I find it looks particularly well planted alongside 'Golden Celebration', which has taller but otherwise similar growth with golden-yellow flowers.

Variety AUSKEPPY | US plant patent applied for | Introduced 2001

Janet

After we had spent fifty years developing roses with flowers of the Old Rose formation, 'Janet' was the first variety we introduced with the bud shape of the Hybrid Tea. It forms a large shrub of sweeping, arching growth and produces flowers which start as pretty, pointed buds and develop into a nice rosette shape, giving us the best of both worlds. The flowers are held on long stems that bend forward to meet you. Their colour is a mixture of pink flushed with yellow, the outer petals being light yellow and paling almost to white with age. This rose has a lovely, strong, pure Tea Rose fragrance. Its health is excellent.

Named after Janet, in her memory; she had a life-long love of roses.

AS A CLIMBER
This rose makes a good climber reaching 2.4m / 8ft.

Variety AUSPISHUS | US plant patent applied for | Introduced 2003

Jubilee Celebration

This is a superb variety and one of the best in this group. Its growth is strong and unusually graceful, arching and spreading widely to form a broad shrub. Its main branches are long and bend forward to provide large flowers in small sprays. They begin as well-filled rosettes, developing into rather dome-shaped flowers with time. It is difficult to describe their colour, partly because it changes so rapidly, but it could be said to be a dusky pink with a tinge of yellow on the reverse of the petals, providing a most unusual and pleasing two-tone effect. The fragrance is typical of this group, being strong and fruity, with a hint of fresh lemon and raspberry. It has characteristically glossy foliage. Like most of our more recent varieties, it is very healthy—in fact, it is a very reliable rose in every way.

Named to mark the Golden Jubilee of Her Majesty Queen Elizabeth II.

Variety AUSHUNTER | US plant patent applied for | Introduced 2002

Pat Austin

'Pat Austin' provides us with a wonderful new colour in English Roses: a mixture of bright copper on the inside and pale copper-yellow on the outside, creating a brilliant effect without being in any way gaudy. It may be somewhat paler later in the season. The flowers are a little more dish-shaped than cupped, delightfully loose-petalled and held nodding on long stems. They have a strong Tea Rose fragrance with a warm, sensuous background. This rose forms a large shrub with elegantly arching growth that shows off the flowers to perfection. The foliage is attractively glossy, though a little less disease-resistant than we would have liked—but some allowances have to be made for a new colour.

I named this beautiful rose after my wife.

Variety AUSMUM | US Patent no. 9527 | Introduced 1995

St. Alban

The buds of 'St. Alban' are globular in shape and pale lemon in colour. The outer petals gradually peel back to reveal a rather deeper yellow within, the full flower forming a beautiful cup with many petals that vary from soft yellow at the centre to almost white on the outer layers. The flower, which slightly hangs its head, has a certain informality but retains its shape throughout. It has a pleasing, fresh scent that is hard to define, but is described by our expert as similar to the fragrance we experience when we walk into a florist's shop. The growth is of medium size, broad, arching, graceful and rather lax in character, with quite small, light green, disease-resistant foliage. 'St. Alban' is a beautiful rose—but not particularly happy in wet weather, when the flowers sometimes fail to open.

Named in honour of the Royal National Rose Society, founded in 1876 and based in St. Albans, Hertfordshire. The first society of its kind, over the years it has done much good work on behalf of the rose.

Variety AUSCHESTNUT | US plant patent applied for | Introduced 2003

St. Cecilia

This is an elegant and attractive rose with flowers that turn to look at you in the most appealing manner. They are chalice-shaped rather than cupped. Their colour is pale buff-apricot, eventually becoming a very pale pink. In autumn they are inclined to be pale throughout. The foliage is large and plentiful, similar to that of a Hybrid Tea. The fragrance is very strong and unusual, of a myrrh character combined with lemon and almond blossom. 'St. Cecilia' forms a medium-sized shrub of upright growth.

Named after the patron saint of music and musicians.

Variety AUSMIT | US Patent no. 8157 | Introduced 1987

Summer Song

The most striking feature of this interesting rose is its colour, which might best be described as burnt orange—a shade that reminds me of tree peonies in Oriental paintings and which is entirely new to English Roses and seldom seen in any rose. It holds this colour, almost without fading, to the end. The flower is beautifully formed, of medium size, cup-shaped and well filled with petals. It has what we call a 'florist's shop' fragrance, a little like chrysanthemum leaves with a hint of tea. The disease-resistance is good. 'Summer Song' is something of a mongrel, having ancestors from other groups, but the overall influence is from the Leander Group.

Variety AUSTANGO | Introduced 2005

Teasing Georgia

'Teasing Georgia' is a superb example of the Leander Group at its best. The flowers are of a lovely, unfading, soft yellow. They open to form a perfect, slightly incurved, rosette-shaped flower of great delicacy. They make a fine display throughout the summer. With its soft colouring and rather pale green leaves, this variety has a suggestion of the English Musk Rose about it, but its breeding and strong growth indicate that it should be placed in this group. It has a particularly fine Tea Rose fragrance, for which it was awarded the Henry Edland Medal in the year 2000 for the best scented variety in the Royal National Rose Society's trials. It has won many other awards in trials in the UK and around the world.

Named for Mr Ulrich Meyer, after his wife Georgia; both are German media personalities.

AS A CLIMBER

'Teasing Georgia' is equally good as a climber. It will cover a large area and yet produce as good a display in the autumn as it has in the summer.

Variety AUSBAKER | US plant patent applied for | Introduced 1998

The Alnwick Rose

An exceptionally strong and healthy rose of medium to tall growth. The flowers are medium sized and pleasing at all stages, being deeply cupped and incurved at first, gradually opening a little to show attractively crinkled petals within—with the slightest tinge of yellow at the centre. They have a rich Old Rose fragrance with a hint of raspberry. The plentiful, healthy, polished foliage is typical of the group. 'The Alnwick Rose' is an unusually reliable rose that is easy to grow.

Named for the Duchess of Northumberland, who has created one of the most impressive large gardens of our time, which includes a magnificent display of English Roses.

Variety AUSGRAB | Introduced 2001

The Ingenious Mr Fairchild

This rose owes its beauty to the fact that its deeply cupped flowers are filled with upstanding petals, each twisting and turning and yet perfectly placed to give a most pleasing effect. The colour of the flowers is deep pink touched with lilac, the outside of the petals being a paler shade, resulting in an attractive two-tone appearance. They have a delicious rose fragrance with aspects of raspberry, peach and a hint of mint. This is a medium-sized shrub of bushy growth, building into a shapely mound with the blooms nicely poised on its branches. It is remarkably free of disease.

Named after the title of Michael Leapman's excellent biography of Thomas Fairchild, a London nurseryman and Fellow of the Royal Society, who made the first recorded flower hybrid in Europe in the year 1720. This was a cross between a sweet William and a carnation, which became known as 'Fairchild's mule'.

AS A CLIMBER

We believe that this beautiful rose will make a good climber. Its arrangement and the manner in which the flowers are held makes it ideal for this purpose.

Variety AUSTIJUS | Introduced 2003

Wisley

'Wisley' bears large, well-formed flowers filled with petals that are deep pink on the inside and paler pink on the outside. They have a fruity, rather citrus fragrance with some element of the Old Rose scent. The overall appearance of this rose is suggestive of an English Old Rose Hybrid, but it belongs in this group by breeding. It develops into a rounded, medium-sized shrub. It is a most beautiful rose that is worthy of the famous name it bears.

Named after the Royal Horticultural Society's foremost garden in Surrey, this is one of four roses that commemorate the foundation of the society formed for the improvement and practice of horticulture in 1804 by John Wedgwood, Sir Joseph Paxton, William Forsyth and three others.

Variety AUSINTENSE | US plant patent applied for | Introduced 2004

A Gallery of English Roses

Ambridge Rose

A short, upright rose, this can be used both in a rose bed or in a mixed border. The flowers are cupped at first, opening into a rosette shape. The inside of the petals is apricot pink, fading to a pale apricot pink at the edges. The outside of the petals is the palest pink, tinged with apricot. This is a good, reliable and healthy rose that flowers freely and continually. The foliage is typical of the Leander Group. There is a fine English Rose myrrh fragrance.

Named after the village in BBC Radio 4's The Archers.

AUSWONDER | US Patent no. 8679 | 1990

Charles Darwin

At first sight, this rose is rather like 'Abraham Darby', both roses having large, cupped flowers on a strong-growing shrub, but the colours are quite different. 'Abraham Darby' is a mixture of pink and yellow—leaning sometimes towards pink and sometimes towards yellow—whereas 'Charles Darwin' is an almost pure mustard shade. There is a delicious fragrance, somewhere between a soft floral tea and almost pure lemon.

Named after the British naturalist who was a native of Shropshire.

AUSPEET | US Patent no. 13992 | 2003

Charles Austin

'Charles Austin' was one of the first members of the Leander Group and is still a worthwhile rose. Its growth is strong and upright with large, glossy leaves. The flowers are large, rosette-shaped and of a rich apricot-yellow, fading a little with age. Its growth is tall and may require cutting back. It will thrive under any reasonable conditions. There is a strong, fruity fragrance.

Named after my father.

AUSFATHER | 1973

Christopher Marlowe

A new colour combination in English Roses, 'Christopher Marlowe' has inner petals of an intense orange-pink inside and golden yellow tinged with pink on the outside, with a yellow button eye. This provides an interesting two-tone effect to the flower as a whole. The flowers, which are of medium size, are cupped in the bud, becoming a rosette shape, the outer petals eventually falling back. They are borne with excellent continuity. It forms an attractive little shrub—spreading and closely packed—with scrawly, intertwining growth. The foliage is medium green and polished with small, shiny leaves similar to that of the R. wichurana *ramblers. It is very healthy.*

Named after the playwright and contemporary of William Shakespeare.

AUSJUMP | US Patent no. 14,943 | 2002

Leander

This is one of the first roses of our Leander Group and the one after which the group was named. It is not the most beautiful of the English Roses but it forms an excellent, very vigorous and colourful shrub which thrives almost anywhere—a fact that makes it very useful. Its flowers are small to medium in size, of rosette shape and held in sprays. Their colour is deep apricot. The foliage is shiny, disease-resistant and of modern appearance.

Named after the legendary Greek lover.

AUSLEA | 1982

Morning Mist

A very large, bushy shrub, 'Morning Mist' is larger than any other English Rose. Its flowers are of medium size, single and pale coppery pink on the inside of the petals, becoming yellow at the centre, and the outside of the petals are a deeper pink. These are beautifully set off by prominent red stamens with yellow anthers. There is a light fragrance of clove and musk. This is a useful garden shrub that is dainty both in flower and growth.

AUSFIRE | 1996

Lilian Austin

Like 'Leander', this rose is of rather modern appearance and is therefore not quite typical of an English Rose. The flowers are semi-double, sometimes almost double. They are salmon-pink in colour, shading to yellow at the centre with yellow stamens—eventually opening wide and flat with wavy petals. They have a strong, fruity fragrance. A medium-sized shrub of broad, graceful growth with glossy, dark green foliage, it has excellent health.

Named after my mother.

AUSMOUND | 1973

William Morris

This is one of a series of excellent roses which includes 'Crown Princess Margareta', 'Geoff Hamilton' and 'James Galway', all of which are very reliable, have a similar habit of growth and have very much the same breeding. The flowers of 'William Morris' are cupped and of medium depth, well filled with petals. Their colour is a lovely, soft apricot-pink, which is a little paler on the outside of the petals. The growth is tall and slightly arching with light green, 'modern' leaves and few thorns. This is a vigorous rose with excellent disease-resistance that forms a large, well-rounded shrub.

Named after the artist, designer and writer to mark the centenary of the University of East London.

AS A CLIMBER *Grown as a climber, 'William Morris' will achieve about 2.4m / 8ft in height.*

AUSWILL | 1998

A Gallery of English Roses

GROUP

III

THE
ENGLISH
MUSK
ROSES

OPPOSITE The English Musk 'Crocus Rose', shows how roses in the English Musk Group are rather lighter in weight and more delicate in appearance—both in flower and leaf—than the Old Rose Hybrids or the Leander Group.

FOR A LONG TIME there has been a number of roses of various types that are in part related to the Musk Roses. These include the old Noisette Roses and the more recent Hybrid Musk Roses, as well as some other varieties of no particular group. These roses have their own special appeal and it occurred to me that by hybridising them with English Roses of the Old Rose group, we might obtain some varieties that share this appeal—and perhaps build on this.

'Iceberg' (bred by Kordes, 1958) is usually classified as a Floribunda but is really a Hybrid Musk Rose, being bred from a cross between the Hybrid Musk 'Robin Hood' and a Hybrid Tea called 'Virgo'. In recent years 'Iceberg' has been deservedly one of the most widely planted of all roses. It has excellent bushy, rather shrubby growth, is very vigorous and repeat-flowers admirably. The flowers are white, but, if you observe them closely, there is sometimes a tinge of blush pink, especially later in the year. It seemed to me that if 'Iceberg' could be crossed with larger-flowered English Roses, we might produce a rose with a different kind of beauty from that of our other roses, drawing upon the particular character of its Musk Rose ancestors.

More recently we have used the Noisette Roses, which are even closer to Musk Roses in their ancestry, to reinforce our group of English Musks still further. The Noisettes are notable not only for the delicate charm of their flowers, but also for their vigour and remarkable ability to resist disease.

It is not easy to describe the English Musks as a group, although they are immediately recognisable. Their growth tends to be of a paler green and slender and smooth, both in branch and leaf. The flowers have a delicacy of appearance that gives them their own particular beauty. They are usually of exquisite formation. The whole effect is of lightness and charm. Their colours range from palest blush to full pink and there are some excellent yellow varieties. All in all, our Musk Rose Hybrids have flowers with much of the beauty of the Noisette Roses, to which they are so closely related.

I wish that I could say that we had caught the Musk Rose fragrance in the English Musks but, with a few exceptions, we have not—although, oddly enough, we do find this scent in some varieties of other groups. The problem is that most English Roses, like the Old Roses, have fully double flowers and the Musk scent is provided by the stamens, not the petals. And, of course, if you have many petals you can have only a few stamens. Nonetheless, the English Musks do have a variety of fragrances, most of which are strong.

A Gallery of English Roses

Anne Boleyn

This rose of low, spreading growth builds up into a neatly mounded shrub, with soft green foliage that is rather more polished than is usual in this group. The flowers are of cupped rosette shape, the petals arranged in perfect symmetry, with the suggestion of a button eye. Their colour is a soft, warm pink, deepening only slightly towards the centre. They nod nicely on the stem. The whole impression of this rose is of neatness and a pleasing freshness, both in flower and leaf. It is reliable and easy to grow, but unfortunately there is no more than a slight fragrance. It forms a beautiful, highly reliable shrub of small size.

Named after the second of King Henry VIII's six wives.

Variety AUSSECRET | US plant patent applied for | Introduced 1999

Charlotte

This beautiful rose is closely related to 'Graham Thomas', but it is a slightly softer yellow and in some ways is a better variety. The colour is easier to place with other roses. The flowers are beautifully cupped, later becoming rosette-shaped with numerous small petals and a button eye. It has bushy and rather upright growth with good, pale green foliage. It forms a small shrub with good growth. It has a strong and pleasing Tea Rose fragrance.

I have a granddaughter named Charlotte and although this rose was named before her arrival, I have dedicated it to her.

Variety AUSPOLY | US Patent no. 9008 | Introduced 1993

Crocus Rose

For the excellence of its low, spreading, arching growth, this rose is hard to beat. It mounds up gradually into a densely packed shrub of exceptional vigour. The flowers are a little larger than medium size, of a cupped, rosette shape and dotted over the whole shrub. They vary from palest lemon at first to pure white as they develop. Held in small sprays, they mingle with the foliage to excellent effect. They have a light Tea Rose fragrance. There is a slight tendency to mildew early in the year, but this can easily be controlled by a single spraying as soon as it appears—there is unlikely to be a problem later on. We tend to classify this rose with the white-flowered varieties since these are so scarce at the present time.

Named for the Crocus Trust, which helps those affected by colorectal cancer.

Variety AUSQUEST | US Patent no. 14092 | Introduced 2000

Francine Austin

This is the only spray-flowered English Rose, the result of a cross between a Multiflora Rose and a Noisette Rose. It bears dainty sprays of small, glistening white pompon flowers, each perfectly formed and placed nicely apart from its neighbour. These are produced in great abundance on long, arching stems, forming an elegant shrub. Unlike the flowers of many spray-flowered varieties, they are sweetly scented with a mixture of Old Rose and Musk. 'Francine Austin' forms a medium to large shrub. Although the growth is compact at first, it seems to restrict itself to a few thick branches as the years go by. To avoid this it is necessary, from time to time, to cut hard back some of the thick, old stems to encourage new growth.

Named after the wife of my son, David J. C. Austin, who runs the Nursery with me.

AS A CLIMBER
With a little encouragement this rose makes a satisfactory climber, particularly when planted against a wall.

Variety AUSRAM | US Patent no. 8156 | Introduced 1988

Graham Thomas

This is probably the best-known of all the English Roses, and one of the most widely grown roses in the world. It owes its popularity in part to its famous name, but not a little to the extraordinary richness of its yellow flowers and their delicious fragrance. The flowers are cupped in shape and have what Robert Calkin has described as 'a fresh Tea Rose fragrance with a cool violet character'. It was awarded the Henry Edland Medal for Fragrance in 2000.

The growth is perhaps less than perfect, being rather upright and un-shapely, but this problem can be overcome by planting it in groups of three or more. The foliage is attractive and typical of an English Musk Rose, being smooth and of a pleasing shade of pale green. It forms a medium-sized shrub. 'Graham Thomas' might be described as no longer in the top flight of English Roses, although it is a valuable variety, not least for its colour, which is unique among English Roses and rare among other roses. It was awarded the James Mason Award by the Royal National Rose Society in 2000.

Named after one of the great figures in British horticulture in the latter half of the twentieth century. Graham was a frequent visitor to our Nursery and gave us much encouragement and advice. When he died in 2003, we had been friends for over fifty years. He chose this rose himself and always loved it.

AS A CLIMBER

'Graham Thomas' is rather better as a climber than as a shrub; in fact, it is one of the best English Climbing Roses. It can easily reach 3m/10ft on a wall, where it will flower repeatedly from early summer onwards.

Variety AUSMAS | Introduced 1983

Heritage

This is one of our earlier roses and one of the original English Musks. Its flowers are of the softest blush pink, medium-sized and distinctly cupped in shape, with a delicate shell-like quality which gives them a special beauty. They have a wonderful fragrance with overtones of fruit, honey and carnation on a myrrh background. The growth and foliage are typically Musk, the leaves being fine rather than heavy and of a greyish-green colour, while the growth is slender but vigorous and slightly arching. It has one sport, 'Rose-Marie', named after the wife of Paul King of Valderose in Canada, who discovered it. Except for its colour—white—which makes it most useful, as good whites are scarce among English Roses, 'Rose-Marie' has all the characteristics of 'Heritage'.

('Heritage') Variety AUSBLUSH | Introduced 1984
('Rose-Marie') Variety AUSOME | Introduced 2003

Lady Emma Hamilton

This variety bears flowers that are chalice-shaped rather than cupped. Their colouring is new to English Roses: a rich apricot-orange on the inside of the petals and flushed with yellow on the outside—the balance changing as more of the inner petals become exposed and varying a little according to the season. There is a strong fragrance with hints of citrus and other fruits. The growth is upright but quite broad, with clean, polished foliage and broad leaflets, resisting disease well. There is a little of the Leander Group in its parentage, but on the whole it fits in best here with the English Musks.

Named after the mistress of Lord Nelson (1765–1815).

Variety AUSBROTHER | Introduced 2005

A Gallery of English Roses

Molineux

This is not a typical member of the English Musk Roses, being closer to the old Tea Roses in its breeding. However, for anyone who requires a good English Rose for planting in a formal rose bed, this variety is one of the very best. It is not very shrubby, being of short and even growth, rather like a Hybrid Tea or Floribunda. It has an extraordinary ability to repeat-flower throughout the season. It is also very reliable and has little disease. The medium-sized, rich yellow flowers form neat rosettes which have a beautiful, characteristic Tea Rose fragrance with a musky background. If 'Molineux' has a weakness, it is that it lacks something of the Old Rose charm we look for in English Roses. It is the least hardy of our roses: in a very cold English winter it may be cut back by the frost, so I would not recommend it for gardens in cold climates.

In 2000 the Royal National Rose Society awarded 'Molineux' a Gold Medal, the President's Trophy for the best rose of the year and the Henry Edland Medal for the best scented rose of the year.

Named for Sir Jack Hayward, former Chairman of Wolverhampton Wanderers Football Club, 'Molineux' being the name of their stadium. Indeed, it must be the only rose named after a football ground.

Variety AUSMOL | US Patent no. 9524 | Introduced 1994

Mortimer Sackler

'Mortimer Sackler' forms a large shrub that requires plenty of space. It sends up tall shoots bearing medium-sized flowers well above the foliage. These are of almost Hybrid Tea shape at first, opening into attractively informal, little cupped flowers of purest soft pink. They have a lovely Old Rose fragrance. It is best pruned in moderation, when it will be ideal for a position towards the back of the border. Excellent health.

Named by Theresa Sackler for her husband.

AS A CLIMBER
This rose may also be used as a climber.

Variety AUSORTS | Introduced 2002

Pegasus

'Pegasus' differs from other English Roses in that it is back-crossed to an early Hybrid Tea for a second time. It also has strong Noisette connections. In appearance it is not unlike the Hybrid Musk 'Buff Beauty'. Its stems are smooth and almost entirely without thorns; the flowers are of a rosette formation and of an attractive apricot-buff colour. Its petals are thick and long-lasting, making it a good cut rose. The fragrance is of the Tea Rose type and particularly strong. The growth is attractively arching, with the flowers drooping slightly on the stem. It has strong foliage with good disease-resistance, being rather like a Tea Rose in this respect.

Named after the winged horse of Greek mythology.

Variety AUSMOON | US Patent no. 9705 | Introduced 1995

Queen of Sweden

This is a rose of unusual freshness and charm. The flowers are beautiful at all stages. They begin as pretty little buds, opening to half-enclosed cups and eventually become slightly incurved rosettes. They have a strict formality throughout, which seems to add to their beauty. Their colour is soft pink, with hints of apricot as the flowers age. The growth is quite short and rather upright, yet bushy, with small Musk Rose foliage that is exceptionally free of disease. It is an excellent rose for arrangement in the house, the flowers lasting several days in water. There is a light myrrh fragrance.

We were both honoured and delighted to be asked to name this rose to celebrate the 350th anniversary of the Treaty of Friendship and Commerce between Queen Christina of Sweden and Oliver Cromwell of Great Britain in 1654.

Variety AUSTIGER | US plant patent applied for | Introduced 2004

Scepter'd Isle

If you require a rose of short stature for a position at the front of a border—or perhaps for a rose bed—this variety would be a good choice. It forms a quite short, bushy shrub that flowers very freely and repeatedly throughout the summer. It is not unlike a shorter version of 'Heritage', to which it is related. The flowers are similarly cupped and soft pink at the centre, shading to pale pink in the outer petals. The leaves, however, are much closer to those of a Hybrid Tea. It has a powerful fragrance—an outstanding example of the English Rose scent based on the myrrh note first introduced with 'Constance Spry'—and, in fact, was awarded the Royal National Rose Society's Henry Edland Medal for fragrance.

The name comes from John of Gaunt's speech expressing his love for England in Shakespeare's Richard II.

Variety AUSLAND | US Patent no. 10969 | Introduced 1996

The Shepherdess

A bushy, well-rounded shrub of rather upright growth, this has unusually large, light green leaves for an English Musk Rose. The flowers are of medium size and attractively cupped, exposing a few stamens when fully open. Their colour is a soft, blush pink with just a hint of apricot. They have a pleasant fruity fragrance with a suggestion of lemon. A remarkably healthy variety with an exceptional ability to repeat flower, this is a charming little rose.

Variety AUSTWIST | Introduced 2005

Wildeve

'Wildeve' is a particularly robust and healthy rose of arching, branching growth, mounding up into a low shrub that is broader than it is tall, not unlike a ground-cover rose. Its flowers, however, are quite large and very beautiful, of about medium size and perfectly rounded, with the outer petals enclosing small, nicely placed inner petals. The colour is soft pink tending towards salmon, with a tinge of yellow at the base of the petals; the outer petals are almost white with a hint of blush—the light reflecting among the petals to enhance the whole effect. The foliage is quite small for an English Rose, very clean-looking and remarkably free from disease. The scent is quite strong, rather unusual and, in fact, impossible to describe.

Named after the character in Thomas Hardy's The Return of the Native.

Variety AUSBONNY | US plant patent applied for | Introduced 2003

Blythe Spirit

'Blythe Spirit' holds its small, double, pure soft yellow flowers nestling in the leaves. It produces these flowers with great freedom throughout the season, although it is necessary to dead head if they are not to turn into hips and thus restrict further production. There is a light Musk Rose perfume with a hint of myrrh. The growth is full, close and very bushy, of medium height and as much across. The leaves are small and medium green, with full leaflets, and are very resistant to disease.

Named after the play by Noel Coward.

AUSCHOOL | 1999

Comtes de Champagne

This is an important rose for us, since it introduced to the English Roses the open-centred, cup shaped flower. The blooms are perfectly globular and rich yellow at first, soon becoming a paler yellow with a boss of deep yellow stamens at the centre. The growth is wide and bushy, producing its flowers on slender, arching stems. There is a delicious honey and musk fragrance. Healthy and free flowering.

Named for Taittinger Champagne in honour of the crusader and troubadour Thibaud IV, Count of Champagne, who is said to have brought Rosa gallica var. officinalis *to France.*

AUSUFO | 2001

Buttercup

This is not a typical English Rose. It shows very little of the influence of the Old Roses. It is, in fact, an entirely new departure which can only be compared with 'Windflower' among English Roses. I see it as a giant buttercup. It forms a large shrub with medium-sized, loosely formed, semi-double, cup-shaped flowers of a strong, deep yellow with darker stamens and a pronounced Tea Rose fragrance. The flowers are held on thin, upright stems well above the pale green foliage. You might try planting this rose at the back of a border behind other shorter plants, with its fine display of bloom showing up against the sky. A very healthy shrub.

Named after the wild flower.

AUSBAND | 1998

Evelyn

This is a short rose that has very large flowers, sometimes giving it a rather unbalanced appearance. At their best the flowers are very beautiful, opening to form very large, flat rosettes of apricot pink, paling a little with age. On a well grown plant they are very imposing. Their fragrance is superb— a beautiful Old Rose scent but with a sumptuous fruity note reminiscent of fresh peaches and apricots.

Named on behalf of the perfumers, Crabtree & Evelyn.

AUSSAUCER | US Patent no. 8680 | 1991

Jayne Austin

This rose bears some of the most refined blooms to be found among the English Musks. Its colour is apricot-yellow and the petals have a lovely, soft, sheeny texture—a characteristic of the old Noisette Roses which has been handed on to this variety. At its best, the flower is a perfectly formed rosette. The growth is slender and rather upright, and a little too lanky to be ideal. To look its best 'Jayne Austin' needs to be planted in a group of three bushes. It has a beautiful fragrance in the Tea Rose tradition, together with a touch of lilac. Unfortunately, the flowers of this rose are subject to the vagaries of the weather. Properly pruned, it will achieve about medium height.

Named after the wife of my son, James Austin, who is a scientist. They live in Yorkshire with their three children.

AUSBREAK | US Patent no. 8682 | 1990

Marinette

Unusual among English Roses, the flowers of 'Marinette' start as long, pointed buds and open to flat, semi-double flowers, each with a bunch of stamens. They are beautiful at all stages. The colour begins as a rose pink, gradually becoming a mixture of white and yellow at the centre and shading to pink at the edges. The flowers are held on dainty, thin stems on an excellent bushy shrub, providing a butterfly-like effect.

Named after Marina Berry, Vicomtesse Camrose, known to her friends as Marinette.

AUSCAM2001 | 1995

Rose of Picardy

A single-flowered rose, this could equally well find a place in groups other than the English Roses. However, we would not like to be without it. 'Rose of Picardy' is a dainty, bright red, single rose that is more than usually prolific. The flowers are about 75mm / 3in across with contrasting golden stamens. They are followed by numerous red hips which have their own beauty, although they will curb later flowering if they are not removed. So we have a choice of flowers or hips later in the season. There is a light, fruity fragrance. The growth is vigorous, forming a bushy shrub of medium height. This variety can equally well be used as a bedding rose, where it makes a brilliant effect.

Named after the popular song written by Frederick E. Weatherley in 1916.

AUSFUDGE | 2004

Sweet Juliet

'Sweet Juliet' is a tall shrub of rather stiff, upright growth. The flowers are medium sized and of a prettily dished, rosette formation, deep apricot in colour, paling towards the edges. They have a delicious fragrance in the Tea tradition, developing a cool lemon character as the flower matures. The shrub is reliably disease-resistant. For good flowers, it requires pruning to about half its height. Though vigorous, it is best planted in a group of two or three because of its slender, narrow growth.

Named after the heroine of Shakespeare's Romeo & Juliet.

AUSLEAP | US Patent no. 8153 | 1989

I HAVE LONG BEEN a lover of the old Alba Roses, which seem to me to capture the very essence of Old Rose beauty. We first used these roses in our breeding programme without any clear idea as to what might be achieved. The result, however, has been a group of roses of considerable charm and delicacy.

The old Alba Roses date back to the Middle Ages. They were the result, it seems, of chance hybridisation between Gallica Roses and the dog rose of the English countryside, Rosa canina. They were once known as tree roses for their tendency to produce tree-like growth rather than to sucker along the ground as is generally the case in many Old Roses when grown on their own roots. They usually have attractive, grey-green foliage that is similar to that of a Dog Rose in all but colour.

Our English Alba Hybrids were produced by crossing the old Albas with English Old Rose hybrids. The result of these crosses is a group of roses that are, on the whole, close to the original Albas in overall appearance, often with semi-double flowers of an almost wild-rose daintiness. Their colour range is rather limited—some are almost white; others range from blush to pink—as is the case with the original Alba Roses. Rather surprisingly, one variety—'Benjamin Britten'—is a beautiful shade of scarlet: this, we hope, may lead to other colours.

Although the English Alba Hybrids are the least fragrant of the four main groups of English Roses, they all have at least a delicate scent—and fragrance should not be judged entirely for its strength. Unlike their colour, this varies widely and includes Old Rose, myrrh, musk and tea, with no particular scent predominating.

The English Alba hybrids have inherited much of the strength and good health of their original parents. Most are Alba in overall character, a feature that we hope to perpetuate. With their airy, natural growth, they associate easily with other plants in the border. They are also probably the best English Roses for associating with shrubs of other genera. We have developed only a few varieties to date, but expect to increase their numbers over the years.

OPPOSITE 'Windflower' illustrates the unique character of the English Alba Hybrids—light, airy growth and flowers that manage to be informal without losing shape.

A Gallery of English Roses

Ann

This is one of the few single roses we have introduced and I think it is probably the most beautiful. The flowers, though they are not showy, have an appearance of delicacy, almost of frailty, that is most appealing. They are exquisitely poised on elegant stems, which shows off their qualities to maximum effect. Their colour is deep rose pink tinged with yellow at the centre, with an occasional yellow stripe and sometimes an extra petal. There is a lovely boss of golden stamens. The fragrance is delicate but pleasing. 'Ann' forms a small shrub which mixes admirably with other plants towards the front of the border.

Named after Ann Saxby who for many years has grown our roses for the Chelsea Flower Show.

Variety AUSFETE | Introduced 1997

A Gallery of English Roses

Cordelia

'Cordelia' bears charming, loosely double, silky-petalled flowers in a delightful shade of purest rose pink that pales slightly with age. They are prettily cupped and carried in sprays, each bloom paling individually to give a pleasing mixture of shades in the cluster. The opening buds have attractively large, elongated sepals. There is a tendency to produce hips, which are handsome but have to be removed if you wish to enjoy further crops of flowers. The growth is very dense, low and spreading—rather as in 'Scarborough Fair'—and the flowers are held daintily above the foliage on thin stems, the whole adding up to a low, well-rounded shrub. The leaves are small, usually with five leaflets. It is a very healthy, reliable shrub.

Named after the king's daughter in Shakespeare's King Lear.

Variety AUSBOTTLE | Introduced 2000

Scarborough Fair

It is easy to be over-impressed with sheer size of flower in a rose. This variety has a more modest beauty, its flowers having a simple charm that is most appealing. They are quite small, rather informal in shape, semi-double and held in sprays. The petals, as they open, incurve to form a ball, which gradually opens to a cup-shaped flower of the utmost delicacy, displaying a bunch of golden stamens—charming at all stages. 'Scarborough Fair' flowers with remarkable freedom and continuity from early summer until the autumn and is ideal for a position towards the front of the border, where its apple-blossom beauty mixes easily with many other colours. The growth is short, broad and dense, forming an unusually neat and well-rounded little shrub. The foliage is quite small, usually with five leaflets and otherwise similar to that of the old Alba Roses. It has a delightful, fresh 'green' Old Rose fragrance of medium strength, sometimes tending towards musk.

Named after the medieval English folk song, made popular by Simon and Garfunkel.

Variety AUSORAN | Introduced 2003

The Alexandra Rose

A tall, slender, slightly arching shrub of anything up to 2.4m / 8ft in height, bearing quite small and dainty, single flowers of coppery pink, gradually fading to pale pink. These open in succession in many-flowered sprays and have conspicuous golden stamens and long sepals. There is a soft Musk Rose fragrance. 'The Alexandra Rose' has Rosa canina *in its parentage, which probably accounts for its attractive, airy growth. It is hardy and disease-resistant and repeats well. It is best planted in groups of three or more. It is ideal for the back of a border where its slender stems will stand above other plants—and useful wherever a natural effect is required. Named for the Alexandra Rose Day, which helps charities to raise money for various voluntary organisations.*

Variety AUSDAY | Introduced 1992

A Gallery of English Roses

Windflower

A number of discerning people have said that they regard this as one of the most beautiful of the English Roses. Not everyone will agree, but there is no doubt that its flowers have an enchanting, almost wild-rose beauty that is hard to compare with any other garden rose—except, perhaps, 'Buttercup' in the English Musk group. There is a strong Alba influence evident in 'Windflower' with its tall, dainty, rather open growth and fine stems, giving it an appeal that is unique among garden roses. The flowers are light and airy, cupped in shape and soft pink, with the slightest touch of lilac. They are held on thin stems well above the foliage, which is of a dainty, pointed, Rosa canina *appearance. All this gives the shrub a very special beauty that brings softness to a border of English Roses and fits in well with other plants in the mixed border. There is an Old Rose fragrance with a hint of apple and cinnamon.*

We chose to name this rose after the common name for Anemone because it holds its flowers with something of the grace and poise of the windflower.

Variety AUSCROSS | Introduced 1994

GROUP
V

THE CLIMBING ENGLISH ROSES

OPPOSITE 'A Shropshire Lad' is just right for a small arch and will reach a much greater height on a wall. It will remain in flower for most of the summer.

THE VARIOUS GROUPS of English Roses lend themselves well to the development of climbing varieties, particularly those that have Climbing Roses in their ancestry. Most English Climbers were not, in fact, intentionally bred as climbing roses, but were a chance product of other breeding. The step from Shrub Rose to Climbing Rose is only a small one and it is therefore not surprising that good climbers have appeared among the former's progeny from time to time. The larger English Roses, with their often rather lax growth and slightly hanging flowers, make ideal climbing roses. Grown upon a wall, arch or trellis, they show off the beauty of their flowers to excellent effect.

The climbing varieties of English Rose are drawn largely from the Leander and English Musk Rose Groups, although there are one or two from the Old Rose Hybrids. English Climbers are not, in fact, a group in their own right, but are simply English Roses from other groups that have shown their ability to climb. It takes us some time to find out whether or not a new rose is capable of climbing. We often have what appears to be a good shrub and, in the course of time, it turns out to be an even better climber. This happened with the first English Rose, 'Constance Spry', which was introduced as a shrub, but has since proved itself to be even better as a climber.

While it is not always possible to define an English Rose as simply being a climber or a shrub, there are some varieties that can be specifically classified as Climbing Roses and others that are better described as 'Shrub/ Climbers'. There are, in this latter group, some really good climbing roses. In my descriptions of shrub varieties, I have sometimes added notes on how they perform as climbers, and I recommend careful consideration of these before making a choice for your garden. Shrub/Climbers are generally best grown on a wall, as this will have the effect of drawing them up to a greater height. So grown, they will usually be more free-flowering and bloom more continually than most true Climbers of whatever kind.

English Climbers will usually reach heights of 2.4–3m/8–10ft—more on the walls of a house. This is just about right for easy management. In climates warmer than that of the British Isles, English Climbers will often grow to a much greater height. They are quite at home on an east or west wall—and even on a shady wall.

The varieties described below are suitable primarily as climbers, but all will form fine large shrubs and these qualities are also noted.

A Gallery of English Roses

A Shropshire Lad Climbing Leander

This rose has peachy-pink flowers of cupped formation which pale towards the edges. As the flowers age they gradually turn back their outer petals, which become a paler shade. They have a deliciously fruity fragrance in the Tea Rose tradition. The foliage and overall habit are typical of the Leander Group, with glossy, deep green leaves and vigorous growth. This rose manages to combine vigour with two very good crops of flowers a year. It is excellent for growing on a wall and is quite capable of covering an arch, providing this is not too large. 'A Shropshire Lad' is a thoroughly reliable climber which will achieve 2.4–3m/8–10ft in height.

Named after A. E. Housman's A Shropshire Lad. *It is a little unfortunate that we chose this name having already named a rose 'Shropshire Lass', as the two so easily become confused, although they are quite different.*

AS A SHRUB

'A Shropshire Lad' is equally successful grown as a large shrub, when it develops pleasing, arching growth, holding its flowers on nodding stems.

Variety AUSLED | US Patent no. 10607 | Introduced 1996

James Galway Climbing Leander

Perhaps the most striking characteristic of 'James Galway' is the neatness of its flowers. All its numerous petals seem to be perfectly placed, forming a domed flower of a little more than medium size. They are warm pink at the centre and, as is often the case with pink roses, pale gradually towards the edges. They have a delicious Old Rose fragrance. This rose makes a robust climber of medium height: perhaps 2.4m/8ft on a pillar or 3m/10ft on a wall. It is very free from disease and repeats well.

Named after the world-famous flautist, in celebration of his 60th birthday.

Introduced in 2000 at the Chelsea Flower Show, where we had the great pleasure of listening to James Galway play.

AS A SHRUB

This rose is every bit as good as a shrub. The growth is tall and slightly arching, making an excellent border plant.

Variety AUSCRYSTAL | US Patent no. 13918 | Introduced 2000

Malvern Hills Climbing English Musk

This variety, unlike most English Climbers, grows to a considerable height —at least 3.6m/12ft and sometimes much more. It might be classified as a repeat-flowering Rambler and is thus a very important rose, as no more than a very small number of varieties could be so described at present. It bears flowers of about 50mm/2in across, in small and medium-sized clusters. Their colour is buff, becoming pale yellow, and they are in the form of a rosette with a button eye. There is a delightful Musk Rose fragrance. Unlike most English Roses, 'Malvern Hills' is not really suitable for growing as a shrub. The growth is strong but slender with smooth, polished leaves and small leaflets. It is ideal for clothing arches and trelliswork. It might be encouraged to scramble over other shrubs or into a small tree. It has almost perfect disease-resistance and will give minimum trouble. Late in the season this rose has the pleasing habit of sending down long, trailing shoots with flowers at each joint. All in all, it is not quite so free-flowering as we would like, but it has a lot of work to do providing such tall growth and flowering throughout the year.

Named after the beautiful range of hills not far to the south of our Nurseries and once home to the composer, Sir Edward Elgar.

Variety AUSCANARY | Introduced 2000

St. Swithun   Climbing Old Rose Hybrid

This rose is the result of crossing a Noisette with one of our Old Rose Hybrids, with the character of growth leaning a little towards the latter. It forms a tall climber of 3–3.6m/10–12ft when grown on a wall. The flowers are neatly formed in the shape of a flat rosette filled with petals, with a button eye at the centre. Their colour is soft pink with palest pink at the edges and they have a strong myrrh fragrance. At their best the flowers look like saucers packed with small petals. They turn towards you on the branch, just as we would wish in a climbing rose. The foliage is smooth and of an almost greyish green.

Named after St. Swithun, Bishop of Winchester, to commemorate the 900th anniversary of the consecration of Winchester Cathedral.

AS A SHRUB

'St. Swithun' can equally well be grown as a shrub, in which form it shows little or no signs of its inclination to climb. It makes nice, bushy growth and it would be hard to say whether it is better as a climber or a shrub.

Variety AUSWITH | US Patent no. 9010 | Introduced 1993

Snow Goose Climbing English Musk

This rose is not unlike 'Francine Austin', with its white pompon flowers produced in large sprays. 'Snow Goose' is, however, a little different in character in that the sprays are closely packed and the shiny, white flowers have petals of uneven length, giving a daisy-like effect. Like 'Francine Austin', it does not strictly qualify as an English Rose and is perhaps better described as a repeat-flowering Rambler. It is also very much more of a climbing rose than a shrub. It is very strong-growing and exceptionally healthy—showing almost no sign of disease—and may be expected to reach 3m/10ft or more on a wall. There is a sweet Musk Rose fragrance. 'Snow Goose' is a very reliable rose that repeat-flowers well.

AS A SHRUB

With its long, lax growth, 'Snow Goose', performs beautifully as a ground-cover rose and, smothered in bloom, it gives the appearance of freshly fallen snow.

Variety AUSPOM | Introduced 1996

Spirit of Freedom   Climbing Old Rose Hybrid

This is a variety of true Old Rose beauty, whose flowers are quite large and not unlike those of a rather smaller 'Constance Spry'. They are deeply cupped, well filled with petals and slightly dished towards the centre. Their colour is a soft pink which gradually turns to lilac-pink with age. They are beautifully poised on the branch and have a delicious fragrance with hints of myrrh. The foliage tends towards a greyish-green and this seems to set off the flowers to perfection. 'Spirit of Freedom' is a very healthy rose. It takes time to judge the eventual height of a climber, but we estimate that this variety could achieve a height of 2.4–2.75m/8–9ft.

Named for the Freedom Association, which campaigns for the preservation and extension of freedom in Britain.

AS A SHRUB

This rose is almost equally good as a shrub. It forms a fine, bushy shrub if pruned for that purpose.

Variety AUSBITE | US Patent no. 14973 | Introduced 2002

The Generous Gardener · Climbing English Musk

It is becoming clear that this rose is our most important English Climber up to the present time. Its flowers are beautifully formed, their colour being soft pink, shading to palest pink on the outer petals and eventually fading almost to white. They are cup-shaped and, when fully open, expose their stamens to provide an almost water lily-like effect. They nod slightly on the stem, the whole effect being one of utmost delicacy. There is a delicious fragrance with aspects of Old Rose, Musk and myrrh. In spite of the flowers' appearance, this is a strong-growing rose that will quickly make a substantial climber, reaching 3.6m/12ft or more. It has pale, almost greyish-green leaves, typical of the Musk Rose Group. It is highly disease-resistant.

Named to mark the 75th anniversary of the National Gardens Scheme which has made it possible for us to see so many beautiful private gardens, thanks to the generosity of their owners.

AS A SHRUB

'The Generous Gardener' forms a beautiful, large shrub of arching growth.

Variety AUSDRAWN | Introduced 2002

The Pilgrim Climbing English Musk

For many years we regarded this rose as a medium-sized shrub, and in this form it is excellent. We have, however, gradually come to realise that it is even better when encouraged to climb. It is an odd fact that good, small shrubs are often better when grown as climbers. This is particularly true of varieties of the English Musk Group, and 'The Pilgrim' is no exception. It forms an excellent climber of 3m/10ft or more. The flowers are between medium and large size, of a nice, shallow-cupped rosette shape and a medium yellow colour, paling a little towards the edges. They have a softness of character that is most pleasing. Their fragrance is a perfect balance between classic Tea Rose and English myrrh. 'The Pilgrim' has plentiful soft green foliage that complements the flowers ideally.

Named after the pilgrims in Chaucer's The Canterbury Tales.

AS A SHRUB

'The Pilgrim' forms an excellent shrub that can be kept quite short and sends up only the occasional long shoot towards the end of the season. These should be cut back to retain its shape.

Variety AUSWALKER | US Patent no. 8678 | Introduced 1991

GROUP

VI

THE ENGLISH CUT-FLOWER ROSES

THESE ARE A NEW DEPARTURE in English Roses, having been bred especially for the cut-flower trade. At the time of writing, we have four varieties. 'Olivia Austin', named after my granddaughter, is a pure rose-pink with a delicious, fruity fragrance. It is related to the Musk Roses and has a typical Old Rose charm. Its cropping capacity is very good. 'Juliet' bears flowers of a rich, creamy apricot-yellow which gradually softens with time. Of all the four cut-flower varieties, this has flowers of the highest quality—it is of true Old Rose character with a waxy sheen which it owes to its Musk Rose parentage. It has a light Tea Rose fragrance. Its cropping ability is excellent. 'Oberon' has flowers that could be compared to those of an old Gallica Rose. They are of a slightly incurved rosette shape and a rich purple-mauve, with a strong Old Rose fragrance. This variety has good cropping ability. 'Portia' is not quite as consistent as the other three varieties, but is superb at its best and is widely used for weddings. It bears very full flowers, blush at first, opening to pure white, with a strong myrrh fragrance.

All the English Cut-Flower Roses have a certain crisp beauty that is partly the result of their breeding—a mixture of the English garden rose with cut-flower Hybrid Teas—and partly the result of their having been grown under the protection of glass. The four current varieties are rather similar in character, being in the form of perfect rosettes, but we aim to widen their range to something more like that of the English Roses for the garden. All these roses have an exceptional cropping capacity and very definite fragrance. A bowl of English Cut-Flower Roses will brighten up the home in winter and remind us of the glories to come in our gardens.

English Cut-Flower Roses are sold in some of the more sophisticated flower shops, and we expect them to become available at most florists. They are especially popular for weddings and special occasions. At present, we have no plans to offer plants to gardeners as we doubt whether many would want to grow roses in their greenhouses.

OPPOSITE A December arrangement of 'Juliet' and 'Portia' with the hips and foliage of *Rosa* 'Frances E. Lester'. The English Cut-Flower Roses have much of the charm and character of the garden varieties of English Roses.

'Olivia Austin'	Variety AUSNOTICE	US plant patent applied for	Introduced 2004
'Juliet'	Variety AUSJAME	US plant patent applied for	Introduced 2004
'Oberon'	Variety AUSVISIT	US plant patent applied for	Introduced 2004
'Portia'	Variety AUSNEIL	US plant patent applied for	Introduced 2004

SOME EARLIER ENGLISH ROSES

Over the last half-century that we have been breeding and introducing new varieties of the English Roses, it is inevitable that some of these would be superseded and that better ones would take their place. Many of the roses that have been removed from our main list have their virtues, even if they also have certain weaknesses.

Some may be beautiful but lacking in health and vigour; others have a tendency to disease and so on. Other varieties, some of them quite recent, have failed to live up to our expectations having been sent out into the world. One or two have been excluded from our main list due to lack of space in this book. But there is also the fact that some of these roses do extremely well in climates other than Britain and, if for no other reason than this, we do not like to lose them. There are also people who collect all our roses.

For all these reasons and also for the fact that many of them are 'old friends', we are loath to let them go. All the roses described below are grown somewhere in the world—and nearly all are available at our Nursery.

Bow Bells

A charming rose that flowers particularly freely, it will form a good bushy shrub of medium height. The fragrant flowers are quite rounded and of a pure rose pink.

AUSBELLS | 1991

Bredon

A short shrub with small, buff yellow flowers of full petalled, rosette formation. It is quick to repeat flower but benefits from generous treatment.

AUSBRED | 1984

Canterbury

Very large, pure pink, almost single flowers with beautiful petals of a silky quality. A short, rounded shrub, it is very lovely but unfortunately lacks vigour. One of the first English Roses. Fragrant.

AUSBURY | 1969

Charmian

Heavy, full petalled flowers of deep, strong pink on a shrub of medium height with arching growth. Unfortunately not too healthy. Very strong Old Rose fragrance.

AUSMIAN | 1982

Chaucer

Large, deeply cupped, rose pink flowers of true Old Rose character but with a strong scent of myrrh. Bushy, upright growth of medium height though not particularly full or strong.

AUSCER | US Patent no. 0000 | 1970

Chaucer

Large, very full petalled flowers of soft pink, paling to blush white at the edges. Unfortunately the flowers tend to spottle to a deep pink in the rain. A tall upright shrub. Fragrant.

AUSLIGHT | 1986

Cressida

A particularly strong, prickly shrub with apricot-pink flowers of full-petalled rather informal formation. Unfortunately not too healthy and a little coarse. Strong myrrh fragrance.

AUSCRESS | 1983

Dr Jackson

A very large, rounded shrub with very pure scarlet-crimson, single blooms. It only flowers once but is followed by a superb crop of large red hips. A wonderful spectacle at both times.

AUSDOCTOR | 1987

Ellen

Large, deeply cupped flowers of soft apricot, deeper in the centre. A strong, bushy shrub of medium height. Strong fragrance.

AUSCUP | 1984

Emmanuel

A very free flowering variety with large, full-petalled flowers in the most delicate blush and apricot tints. Delicious, fruity fragrance. It forms a bushy shrub of medium height, although, unfortunately, rather susceptible to blackspot.

AUSUEL | 1985

English Elegance

Large, loose-petalled flowers with petals of a wide range of colours from pink and salmon to copper, a wonderful mix. A tall shrub that is rather sparse flowering but seems to do better in warmer climates.

AUSLEAF | US Patent no. 7557 | 1986

English Garden

Beautiful, flat and perfectly formed flowers of soft apricot-yellow at the centre, paling towards the outside. A fairly short shrub of upright habit, it would be particularly good as a bedding rose but benefits from regular spraying. Light fragrance.

AUSBUFF | US Patent no. 7214 | 1986

Fair Bianca

A pure white rose with blooms of true Old Rose character. Short, upright, although rather sparse growth that benefits from being grown in a Mediterranean-type climate. Strong myrrh fragrance.

AUSCA | 1982

Financial Times Centenary

Large, cupped blooms of a wonderfully clear, deep pink. A medium sized, upright shrub that benefits from some extra care. Powerful fragrance.

AUSFIN | US Patent no. 8142 | 1988

Fisherman's Friend

Large and beautiful, full-petalled flowers varying from garnet to deep cerise crimson. A very thorny upright shrub of medium height that benefits from regular spraying. Strong Old Rose fragrance.

AUSCHILD | 1987

Happy Child

Large, shallowly cupped flowers of an exceptionally rich, deep yellow. The leaves are highly polished but not very resistant to blackspot, and so this rose does better in drier climates. Forms a bushy shrub of medium height. Delicious Tea Rose fragrance.

AUSCOMP | US Patent no. 9007 | 1993

Heavenly Rosalind

A very free-flowering variety with medium-sized, single flowers of soft pink. A tall shrub giving a lovely wild rose effect, although it will benefit from occasional sprays.

AUSMASH | 1995

Hilda Murrell

Large, almost perfectly formed flowers of the richest, purest, shining pink. The growth is strong and upright, producing a wonderful show of flowers in early summer with a smaller crop later. Rich and powerful fragrance.

AUSMURR | 1984

Jacquenetta

A very free- and long- flowering large shrub with large, semi-double flowers of blush pink tending towards apricot.

1983

Lilac Rose

Fine, large blooms of lilac pink on a short, bushy shrub. Needs regular sprays to keep it healthy. Very strong myrrh fragrance.

AUSLILAC | US Patent no. 8837 | 1990

Lucetta

Large, semi-double flowers of soft blush pink fading almost to white. A shrub of medium height with strong, arching, healthy growth. Fragrant.

AUSEMI | 1983

Ludlow Castle

A charming rose with flowers of pristine beauty. Their colour is a gentle apricot-blush, which gradually pales to a soft apricot-white. Unfortunately this variety, though very beautiful, is weak and subject to disease. Fresh Tea Rose fragrance.

AUSRACE | US Patent no. 13299 | 2000

Mary Webb

Giant, almost peony-like flowers of pale lemon yellow that benefit from dry summers to open properly. A robust, bushy, upright shrub of medium height. Fragrant.

AUSWEBB | 1983

Mayor of Casterbridge

Close to an Old Rose, both in flower and leaf. Cup-shaped flowers of a pure soft pink with a lighter reverse. Vigorous, upright growth. A light, fruity Old Rose fragrance.

AUSBRID | 1996

Moonbeam

The medium sized flowers of glowing white are produced with exceptional freedom. A vigorous shrub of medium height.

AUSBEAM | 1983

Othello

Very large, deeply cupped and many petalled blooms of dark crimson turning to shades of purple and mauve. A very robust and thorny shrub of medium height. Strong Old Rose fragrance.

AUSLO | US Patent no. 7212 | 1986

Peach Blossom

Medium-sized, semi-double flowers of sheeny, almost transparent delicacy. A medium-sized shrub that flowers freely and repeats well but will produce a good crop of hips if not dead headed. Fragrant.

AUSBLOSSOM | 1990

Perdita

Rosette-shaped flowers of a delicate apricot blush. Strong myrrh fragrance. A beautiful rose at its best, but no longer quite up to the necessary standard.

AUSPERD | 1983

Pretty Jessica

A short and very compact shrub with perfectly formed flowers of warm, rich pink. It repeat-flowers very well but needs regular spraying. Strongly fragrant.

AUSJESS | 1983

Radio Times

Classic Old Rose shaped blooms of a beautiful rosette shape and a rich, clear pink. A short, bushy shrub that is, unfortunately, not too healthy but grows very well in warmer drier climates. Very strong Old Rose fragrance.

AUSSAL | US Patent no. 9525 | 1994

Sir Clough

The beautiful crimson buds open to large, semi-double flowers of a bright and very deep pink. It makes a tall although a slightly sparse shrub.

AUSCLOUGH | 1983

Sir Edward Elgar

Large, extremely full-petalled flowers of a beautiful cerise crimson. A bushy slightly sparse shrub that benefits from a drier, warmer climate, which helps with health and makes the fragrance stronger.

AUSPRIMA | US Patent no. 8670 | 1992

Sir Walter Raleigh

Exceptionally large, double flowers with open centres and of a clear warm pink. It forms an attractive, rounded shrub of medium height. Strong and pleasing fragrance.

AUSPRY | US Patent no. 7213 | 1985

Swan

A very strong, upright shrub with very wide, flat, full-petalled, white flowers that last very well when cut. It favours a dry climate as the flowers can spottle badly in the rain. Fragrant.

AUSWHITE | US Patent no. 7564 | 1987

Tamora

A short, upright, bushy plant producing large, full-petalled, deeply cupped flowers in a pleasing shade of apricot. Particularly successful in warmer climates where it is an excellent bedding rose. Very strong and delicious myrrh scent.

AUSTAMORA | 1983

The Nun

An unusual rose with deeply cupped, semi-double white flowers of almost tulip formation. An upright shrub of medium height, it flowers freely but the stems are rather sparse and disease resistance is poor. Slight scent.

AUSNUN | 1987

The Prince

Beautifully formed flowers of the deepest, velvet crimson, turning quickly to a rich royal purple. A short, rather weak shrub that benefits from generous treatment. More successful in warmer areas.

AUSVELVET | US Patent no. 8813 | 1990

The Prioress

Open, semi-double flowers of white tinted with blush. It has vigorous upright growth and is particularly good in late summer. Fragrant.

1969

The Reeve

Dusky, deep pink flowers of globular shape. The stems are rather prickly and arch over, possibly too much. Strong Old Rose fragrance.

AUSREEVE | 1979

The Squire

Very beautiful, full-petalled cupped blooms of deep crimson. A short, upright shrub that is a little sparse benefiting from a warm climate and generous treatment. Strong Old Rose fragrance.

AUSQUIRE | 1977

Tradescant

Beautifully formed, rosette-shaped flowers of rich wine crimson turning to the deepest, richest purple. In temperate climates like the UK it is a little on the weak side but in Mediterranean climates it grows into a large shrub or short climber. Strong Old Rose fragrance.

AUSDIR | US Patent no. 9009 | 1993

Troilus

Very full-petalled, cup-shaped flowers of a lovely honey buff shade. It is a sturdy, upright shrub of medium height, but it prefers growing in a climate with dry summers as the flowers are susceptible to balling. A sweet honey fragrance.

AUSOIL | 1983

Warwick Castle

Beautifully formed blooms of glowing pink in a perfect flat rosette shape. It has low arching growth that will, given generous treatment, form an attractive rounded shrub. Strong fragrance.

AUSLIAN | 1986

Wife of Bath

One of the first English Roses, it is a short shrub with cupped flowers of rose pink. It tends to suffer from dieback but is amazingly tough and resilient. Strong myrrh fragrance.

AUSWIFE | 1969

Wise Portia

A short bushy rose that can produce some of the most beautiful blooms in lovely shades of purple and mauve. It is hard to beat at its best but it does require generous treatment. Rich fragrance.

AUSPORT | 1982

Yellow Charles Austin

A sport of Charles Austin, identical in every respect except colour which is a pleasing shade of lemon-yellow. A large and very tough and reliable shrub with vigorous upright growth. Strong fruity fragrance.

AUSLING | 1981

PART

THREE

—

ENGLISH

ROSES

IN THE

GARDEN

1

THE
ENGLISH
ROSE AS
A GARDEN
PLANT

THE ENGLISH ROSES were bred with their use in the garden clearly in view. Their flowers have both beauty and fragrance and come in a great variety of forms and sizes, from very large to quite small. But not only this. Their natural shrubby growth—as compared with the squat and angular growth of the Hybrid Teas and Floribundas—makes them ideal for the garden. Their habit may be bushy or elegantly arching. They may vary in height from tall to quite short. Their foliage can be in many forms and colours. The colour of the flowers, while usually in the gentler shades, can vary from white to deep crimson and from pale lemon to rich gold.

With all these variations, there is something among English Roses that will fit into many parts of the garden and with a great variety of other plants. And, of course, like most present-day roses, they flower throughout the summer. They are among the aristocrats of the garden, providing something special around which other plants can be arranged.

The garden rose is very much man's creation, even if this is with a great deal of help from above! The nearer to the house and the nearer to man-made structures, the more at home the roses seem to be. This is certainly true of the English Roses. They are very much in place in the mixed border and in beds and borders surrounding the house. With their rather unruly growth—and the weight of their flowers—English Roses seem ideally suited where they are in marked contrast to hard edges. It is, I think, the contrast between the more formal garden, with its neatly cut edges, and the rampant growth of English Roses that is so effective. Each seems to need the other. Whereas English Roses appreciate the discipline that the formal garden offers, we would not usually plant them in wilder areas, unless we chose one of the English Alba hybrids or another variety of more rampant, wild-rose growth.

The Importance of Planting in Groups

Most people would agree that nearly all plants are more effective and look more natural when planted in groups or drifts of one subject rather than as individual specimens. Flowers in the wild nearly always grow in this way. Numerous different plants produce a dotted, rather random effect and we do not see any particular plant very well. This is true of roses and especially

English Roses which, along with other Shrub Roses, are in one way less-than-ideal garden plants. They are the most beautiful of flowers; they are the most useful of plants that will add something to the garden that no other plant can; but even to rose enthusiasts like me, they have two drawbacks. Because roses produce so many, often heavily petalled flowers and continue to do this throughout the summer, great demands are placed upon the bush.

To assist in this task, roses are usually grown on root stocks to give them extra strength. This means that all the growth comes from one point—the point at which the rose was budded. The result is a shrub that is rather narrow at the base while being broad and full at the top. This does not matter so much if we are dealing with a variety of rose with arching or very wide growth, but it can result in a rather ungainly shrub.

The problem does not end here. The ability to repeat-flower has another important effect on the rose. Nearly all wild roses flower once in early summer. For the re-mainder of the year they use their energy to develop long, strong branches from which they will produce next year's flowers. The repeat-flowering rose of today enjoys no such advantage. It is repeat-flowering only because every branch produces a flower. This often results in a shrub which develops in a rather random way and is not always shapely.

There is a simple answer to the above problems—and this is to plant your roses in close groups of three or more of the same variety. This is far better than three different roses scattered around the garden. Planted about 45cm/18in apart, the bushes will grow together to form a dense but shapely thicket that, to all appearances, is one fine shrub. It is true that this adds to the expense of buying your rose bushes, but the effect in the garden is greatly enhanced and I am sure it is worthwhile. Where space is limited, however, as in a very small garden or the corner of a larger one where plants can be viewed from close quarters, a single rose bush among other plants can be perfectly adequate.

The Mixed Border

English Roses are very well adapted to growing with herbaceous plants and shrubs in the mixed border, and this is where most people wish to have them. In spite of the fact that they are themselves shrubs, they are less suited to a border made up entirely of shrubs. This is because they produce com-paratively highly developed flowers in great quantity. If you do wish to in-clude English Roses with shrubs, it is best to choose larger-growing

OPPOSITE 'The Mayflower' with catmint, *Nepeta* 'Six Hills Giant'.
ABOVE 'Cottage Rose' with herba-ceous geraniums and other plants in a mixed border.

English Roses in the Garden

varieties with natural growth and simple flowers, such as 'Cordelia', 'Mortimer Sackler', 'Scarborough Fair', 'The Alexandra Rose' and 'Windflower'.

My advice on the desirability of planting English Roses in groups applies particularly to the mixed border where they have to compete with other plants which may be more vigorous and invasive. A group of roses of one variety can live happily together, since they are each of equal vigour and no one rose is dominant or will smother the others. It is also very important that other subjects planted around the roses are not too vigorous. A repeat-flowering rose does so much work that it cannot withstand too much competition: even a quite short plant that is invasive can, in the end, overcome or at least reduce the vigour of the rose.

Group planting in the mixed border is equally important for aesthetic reasons. A single rose bush, planted among other subjects, will not only have difficulty in showing itself to maximum effect but can also become lost in a riot of other flowers. English Roses form vigorous shrubs but they do not display their full beauty if squeezed in between other plants. They need to be able to make a strong statement. If they are planted with shorter companion plants around them, this will give them space to display their grace and beauty to the full. That they may do this more completely, it is worthwhile pulling them a little further forward in the border, in relation to the plants around them. In this way they will not be overshadowed by larger plants at the back of the border and they will be able to display both their flowers and their growth to the very best effect. This will have the added advantage of breaking up any tendency to flatness in the border. English Roses are among the more dramatic plants in the garden and it is worth using them as focal points, around which to arrange other plants.

ABOVE Two herbaceous borders with groups of English Roses, creating a lovely informal effect.

OPPOSITE A grass path meandering through a riot of English Roses, including 'Mary Rose', 'Wenlock' and 'Pat Austin' on the left and 'Buttercup' on the right.

The many different varieties of English Roses are notable for their diversity of flower, leaf and growth. This increases as we introduce new strains into the breed and makes them useful for a wider range of positions in the mixed border. Tall, upright growers like 'Charles Austin', 'Graham Thomas' and 'The Pilgrim'—which are sometimes a little ungainly in growth—can be planted behind smaller plants, so that they can look over their heads. Varieties of arching habit, like 'Abraham Darby, 'Crocus Rose',

English Roses in the Garden

'Golden Celebration' and 'Grace', should be given ample space to show off the beauty of their growth. Plant them a little back and encourage them to send their long branches gracefully forward, towards the front of the border. Bushy, twiggy varieties—'Eglantyne', 'Mary Rose', 'Windflower' and many others—will take a more central position in the border. Short, bushy roses, like 'Ann', 'Charity', 'Mary Magdalene' and 'The Mayflower', are useful for the front of the border. There are also roses of low, wide growth, like 'Trevor Griffiths' and 'William Shakespeare 2000', that will also find a place towards the front of the border. We need not, however, worry too much about such matters: much that is good will happen by happy coincidence and, if we are wrong, we can always move either plants or roses and try again. Contrary to what we might expect, even mature roses move well in the autumn and may even improve with such treatment.

A Border of English Roses

If you would like to put together a collection of English Roses, you might consider planting a border of them alone. This is a good way of maintaining a collection, particularly if your garden is limited in size and a complete rose garden is not possible. In such a border, group planting is even more im-portant if it is to have form. It will never be without bloom from late May until the onset of winter. A wide range of varieties can be grown for the various forms of their beauty and—if you wish—their blooms can be cut and enjoyed in the house. As the years go by, new varieties that may appeal to you can be added and less favoured ones taken out. In this way you will have an ever-evolving interest.

In a border intended exclusively for English Roses, their often large and heavy flowers can result in a rather 'blobby' effect unless they are mixed with varieties that have smaller, lighter flowers held in dainty sprays. This has a softening effect and links the border into one harmonious whole. Varieties with smaller flowers include 'Blythe Spirit', 'Buttercup', 'Francine Austin' and 'Mistress Quickly'. There is also a case for including a few roses from other groups. We might also use the Ground Cover Roses, 'Little White Pet' and the

Polyantha Roses, as well as species roses which, with their dainty growth, can be used almost like foliage plants.

You may, of course, take the mingling of English Roses with other roses much further, and include Old Roses. Many people are very fond of the Old Roses—and I share this feeling. The problem is that the most beautiful forms—that is to say, the original varieties—have the disadvantage that they tend to have rather unattractive growth and do not repeat-flower, and the more recent Old Roses, the Portlands, the Bourbons and the Hybrid Perpetuals, that do repeat do so only to a limited extent and are subject to disease. Old Roses in general also have a very restricted colour range. For these reasons alone, it would be worthwhile adding some English Roses to a border of Old Roses. In this way we can have a border of great variety and beauty that will provide pleasure throughout the summer.

Rose Beds

During the twentieth century, when the Hybrid Teas and Floribundas were supreme, it became accepted that the proper place for roses was in formal rose beds; indeed, they were bred for this purpose and I think this still is their ideal home. English Roses—which were bred more especially for the border, like the Old Roses before them and, indeed, Shrub Roses generally—are not usually suitable for such treatment. However, there are a number of English Roses that are short in growth, very branching and exceptionally repeat-flowering which are entirely suitable for rose beds. Some of them will even out-flower many, if not all, Floribundas. This is largely because they were originally bred from vigorous, healthy roses and is due, in some degree, to the hybrid vigour that results when we cross roses of widely differing origins—as is the case with English Roses.

When using Hybrid Teas and Floribundas, with their intense colours, we are usually advised to plant one variety to a bed, otherwise such colours are likely to clash. In the case of English Roses there is no such problem and most varieties can be mixed freely. Only when using the stronger colours do we have to be more careful. A mixture of soft-coloured English Roses can be planted so the colours melt into each other, creating an almost dappled effect. A bed of mixed English Roses, given the right setting and pruned hard, can be a pleasing feature in the garden. Mixing English Roses with brightly coloured Hybrid Teas and Floribundas is unlikely to be successful, however.

BELOW The naturally bushy, spreading growth of English Roses makes them without doubt the best of all roses for standards. Here is 'Graham Thomas'.

English Roses are the finest of all roses for growing as standards, or 'tree roses' as our American friends call them—better than any other kind of rose. You might expect me to make this claim, but I am certain that their large, strong and usually bushy or arching growth makes the English Roses ideal for this purpose. To produce a good head on a single stem requires a strong rose. Some varieties might be too big, but all except the largest form excellent, well-rounded, shapely heads.

Standard roses are valuable for strategic positions in the garden. They have the advantage that we can walk straight up to them, enjoy their blooms and savour their fragrance at close quarters. The traditional idea of two rows of standard roses—set at some distance from each other on each side of a garden path—can be highly satisfactory. A large, well-grown standard rose, standing alone, can make an excellent focal point. On a lawn which is not large enough for a specimen tree, an English Rose grown as a standard can do a similar job, and provide flowers throughout the summer. A variety of somewhat arching growth, such as 'Golden Celebration' or 'Grace', is often the most suitable, since its blooms turn and look at us, rather than staring into the sky.

Standards are useful for creating height. They make excellent centrepieces for small rose gardens—although a piece of sculpture might be even better. While we would not usually put standard roses in a mixed border, they can be effective in borders if planted with their heads standing well above low-growing plants to give a pleasing, two-level effect. Traditional rose beds, particularly as part of a formal rose garden, can sometimes look flat. A standard rose at the centre of each bed will relieve this flatness and add to the effect of formality, so long as the bushes beneath do not come up to meet the head of the standard rose—this only creates a muddled effect. It is important to have very low-growing, hard-pruned bushes surrounding the standards.

Examples of good English Roses for standards are 'Eglantyne' (soft pink), 'Golden Celebration' (rich golden yellow), 'L.D. Braithwaite' (bright crimson), 'Mary Magdalene' (soft apricot pink), 'Mary Rose' (rose pink), 'Molineux' (rich yellow) and 'William Shakespeare 2000' (rich velvety crimson).

Specimen Shrubs

Free-standing English Roses can play much the same role as the standard roses. Large specimen shrubs can be very effective along a drive, for example, or set in a lawn or meadow. They are also appropriate in minimalist gardens.

With English Roses grown as specimen shrubs, it is more than ever important to plant at least three bushes of one variety together to get the best result. These should be planted closely—perhaps 45cm/18in apart—so that they eventually grow into what amounts to a single specimen.

Varieties suitable for growing as specimen shrubs are those that form broad, full shrubs or have wide, perhaps arching growth. The English Rose 'Hyde Hall' is ideal for this purpose. It is related to *Rosa fedtschenkoana*, a large species rose that is naturally repeat-flowering. This has enabled us to breed a large shrub with a remarkable ability to repeat-flower—two characteristics that do not usually go together. Others to choose include 'Abraham Darby' (pink with apricot and yellow), 'A Shropshire Lad' (soft peachy pink), 'Chianti' (deep crimson), 'Crocus Rose' (soft apricot), 'Crown Princess Margareta' (apricot orange), 'Geoff Hamilton' (warm, soft pink), 'Golden Celebration' (rich golden yellow), 'James Galway' (warm pink), 'Leander' (deep apricot), 'Pat Austin' (bright copper), 'St. Swithun' (soft pink), 'Shropshire Lass' (flesh pink), 'Sweet Juliet' (glowing apricot), 'Teasing Georgia' (deep yellow) and 'Tess of the d'Urbervilles' (bright crimson). Careful selection is essential, since varieties with upright growth will be bare at the base, often with all the flowers on top.

Pots and Containers

There is a place for pot-grown roses in every garden, whatever the size. Even people with little more than a patio can grow roses. It is an increasingly popular practice to grow plants in pots and other containers, and English Roses are ideal for this purpose. They often have bushy or spreading growth, which is just what we require in a pot plant standing

alone, and, unlike most other plants, they flower over a long period. Such pots can be arranged on a paved or gravel area to very good effect.

In large gardens pot-grown roses can be placed to bring character and interest to what might otherwise be a plain or boring space, as, for example, a row of six pots would enhance a paved area running along the front of a house. There has been increasing demand at our nursery for the English Roses we supply in large, 46cm/18in pots. Of course, the pots could be smaller and it is possible to purchase bare-rooted bushes and plant them yourself (see page 296).

One advantage of growing roses in containers is that we are given the chance to view them at close quarters, perhaps as only the keenest gardener would see them in the garden proper. The roses become intimate compan-

English Roses in the Garden

ions and we can enjoy every detail of their blooms, their fragrance and their growth. Another bonus is that pots can be moved around and rearranged; thus we can bring a rose that is in full flower to the fore or make extra space for more people and so on. All in all, there seems to be a great future for growing roses in this way.

Roses grown in containers are entirely dependent on us for their livelihood and, obviously, adequate watering and feeding are essential. If roses are allowed to dry out, they will quickly deteriorate.

Almost any English Rose is suitable for growing in pots, the more upright varieties being suitable for smaller pots. We regard the following as particularly suitable for large containers due to their bushy habit of growth. 'Anne Boleyn' (soft, warm pink). 'Comtes de Champagne' (rich yellow), 'Christopher Marlowe' (intense orange-pink), 'Golden Celebration' (rich golden-yellow), 'Grace' (apricot), 'Harlow Carr' (rose-pink), 'John Clare' (deep glowing pink), 'Mary Magdalene' (soft apricot-pink), 'Portmeirion' (rich glowing pink), 'Sophy's Rose' (light crimson) and 'William Shakespeare 2000' (deep crimson).

OPPOSITE English Roses, with their elegant bushy growth, are among the very best roses for growing in pots. Here is 'Benjamin Britten' in full flower.

BELOW English Roses are ideal for informal hedges. This is 'Wild Edric', one of the best varieties for this purpose.

Hedges

There is yet one further way in which the English Rose can make a contribution to the garden and that is as a hedging plant. There is no other—at least, in the temperate world—that can provide so much colour over so long a period as the rose, and English Roses are no exception. While roses do not, of course, make neat, clipped hedges, and you will want most of your hedging to provide a calm, green background, there are places where a colourful hedge is a valuable feature 'within' the garden. Rose hedges may also be useful to divide one part of the garden from another—and English Roses are hard to beat for such a purpose. For houses where the front garden is open to the road, it is a good idea to plant a row of English Roses to mark the point where the road ends and the garden begins. Such a hedge will have the added advantage that it will form an impenetrable, prickly barrier.

It is usual to plant only one kind of rose in a hedge, as a mixture of varieties is likely to result in the strong being victorious, while the weaker varieties slowly fade away. On the other hand, where space is limited, a mixed hedge does give us a chance to increase the number of varieties we grow. If you do plant a mixture, try to select varieties of similar size and strength. For a hedge of low to

medium height, you might consider combining 'Mary Rose' (deep pink), 'Redouté' (soft pink) and 'Winchester Cathedral' (white), since the last two varieties are sports of 'Mary Rose' and, therefore, are all of similar growth.

To achieve a nice dense hedge, plant your bushes closely: not more than 45cm/18in apart. This can be expensive for a long hedge, but your nursery-man will usually make a substantial reduction in price for a large order of one variety.

Rose hedges may vary in height, from low to quite tall. Low hedges are useful more as a dividing line between one part of the garden and another, than as any kind of barrier. Many people will have seen the Gallica Rose 'Officinalis' (now *Rosa gallica* var. *officinalis*) or Rosa Mundi (now *R. gallica* 'Versicolor') grown as low hedges, sometimes edging a large border. There are equally good subjects for this purpose among the English Roses and these will flower repeatedly over a long period. 'Ann', 'The Apothecary's Rose', and 'The Mayflower' are ideal examples. These can be kept low by pruning or clipping.

'The Mayflower' is a particularly suitable rose for hedging, being very twiggy, very repeat-flowering and, as far as we know, completely free of disease. This last is a very important consideration when planting a rose hedge. With attractive dull green foliage and flowers of not too brilliant a colour, this is a rose that can be depended upon not to outshine everything that is planted in front of it. The only problem with 'The Mayflower' is that it is somewhat subject to red spider, which is particularly a problem in the southern United States and areas with similarly hot summers and mild winters. Few gardens in Britain suffer from this pest, but if yours does, do not use this variety. Other low-growing varieties suitable for hedges include 'L.D. Braithwaite', 'Molineux' and 'Scepter'd Isle'.

For a hedge of medium height you might use 'Corvedale', 'Eglantyne', 'Mary Rose', 'Redouté' or 'Winchester Cathedral'.

A tall hedge may require support in the way of a fence or wire netting— it is likely to be smothered with bloom and this could weigh it down so that it loses form. 'Blythe Spirit', 'Buttercup', 'Crown Princess Margareta', 'Teasing Georgia', 'Falstaff', 'Leander', 'Marinette', 'Mistress Quickly', 'Tess of the d'Urbervilles' and 'Windflower' are all suitable tall roses. 'Hyde Hall' is particularly suitable as it forms an impenetrable barrier and flowers with exceptional continuity.

As regards colour, the more muted shades are usually best. I also find that too extreme a mixture is seldom desirable.

Whatever the rose or roses used, with so many growing in close proximity it is likely to be necessary to spray against disease—and do not forget that rose hedges benefit from the same treatment as roses in any other position. Feeding will be amply repaid.

Plant Companions

Whether we are planting English Roses in a mixed border, a rose border, or indeed in a rose garden, the question arises as to how we should place them in relation to other varieties and other plants in order to provide the most pleasing effect. Others, better qualified than myself, have written on the subject in more detail than I have space for here, but I should like to make some general points.

As regards colour, English Roses, with their usually gentle shades, do not usually like too much competition from stronger colours unless, occasionally, you might deliberately wish to create a contrast. English Roses can

ABOVE 'William Shakespeare 2000' in a border with *Capanula lactiflora*.

be broadly divided into two—the yellows and the rest. Yellow English Roses are of very different from roses of other colours and it is interesting to note that they are of very different breeding. Yellow is a 'hot' colour, whereas most English Roses have 'cool' colours. Even red English Roses tend to be cool since they nearly always have a touch of violet or mauve in their make-up. This becomes more evident as the flower ages and makes red roses less of a problem to mix with colours other than yellow.

Hot and cool colours should be mixed only with care. A successful rose garden can be made exclusively of yellows, apricots and flame shades. Equally, it is possible to have a rose garden or border made up of shades of blush pink running through to deep pink—and on to crimson, purple, violet and mauve. If you want to grow a range of English Roses, simply separate the hot from the cool by placing your yellow shades in one area. It is also possible to create a rainbow effect, with one colour running into another, in succession, from one end of the border to the other.

The same rule holds true when mixing English Roses with plants of other genera: cool colours should not be placed too close to hot. However, the rule should not be interpreted too strictly as splashes of hot colour can bring life to a border made up of cool shades. Also, of course, most of us have small gardens and do not have a great deal of choice if we want to have a range of roses.

As regards the question 'What are the best plants to plant with English Roses?', here are one or two points that might be helpful. The first is that many English Roses—like their sisters, the Old Roses and the Hybrid Teas—tend to give a rather 'blobby' effect in the mass. They are beautiful and fragrant, but they are heavy in the individual flower. This in itself is not in any way bad—it is part of their beauty—but it does demand something a little lighter to go with them. In a border of English Roses it is a good idea to lighten the effect by including the occasional rose of a lighter nature. These may be single, semi-double or spray-flowered. There is no shortage of such roses and, as well as those listed on page 247, I suggest the following

ABOVE 'Grace' with 'The Pilgrim' behind, showing their pleasing, natural growth.

OPPOSITE Two borders divided by a stretch of lawn can form a pleasing small rose garden.

English Roses in the Garden

English Roses: 'Heritage', 'Cordelia', 'Corvedale', 'Morning Mist', 'Scarborough Fair', 'Ann', and 'Windflower'. As before, we might add the so-called Ground Cover Roses or some species roses. The latter flower for only a short period, but their foliage is light and airy and they have the added advantage that their dainty flowers are followed by decorative hips. What I am saying is that you should try to vary the weight of the flower to avoid a too-heavy effect. The roses of lighter weight will help to bring the border together into one whole. Of course, in a mixed border of English Roses and herbaceous plants, there is no great difficulty—the range of plants is so great. Even then, I would avoid planting too many large, heavy-flowered plants like peonies together with English Roses. Something altogether lighter in weight is likely to be more suitable. On the other hand, the beautiful spires of delphiniums can be very successful with pink and white roses.

We should not lose sight of the fact that roses flower when many other plants are over and then continue for a very long time. For this reason they are particularly useful and we also avoid the problem of what goes with what. Early in the year the attractive foliage of the English Roses can make a setting for other plants, while later the foliage of other plants makes an

excellent setting for the roses. The same plants will provide a setting for the roses when they are in flower. With a little thought and manipulation, it is possible to use both to advantage.

On the whole, you will know which rose goes best with another rose or with another plant by careful observation. If you have the opportunity to cut a few blooms, both of roses and of other plants, you will soon see which combination pleases you. And if it turns out that you have made a mistake, it is easy to remove the offending rose or plant. As I have said, roses move more easily than most people think, if this is done in autumn.

Despite being well aware of how a garden should be planned, I must confess that our Gardens at Albrighton have no such scheme, having grown a little haphazardly as new varieties were introduced. This does not seem to have affected them adversely, as far as I can tell.

Modern Trends

Those of us who watch gardening programmes on television, or have seen display gardens at flower shows, cannot help being struck by a trend to move gardening on to a school of design in which borders, lawns and plants of all kinds are notable for their scarcity. When gravel, concrete and stone take over, flowers are often incorporated economically and roses are scarcely seen at all. Yet English Roses are ideal in these 'new' gardens: they flower throughout the summer and are not unattractive in spring. Also, well-structured roses show themselves off to maximum effect near man-made structures.

However, we have to accept that there are people who believe that the highest form of gardening dispenses with plants almost completely. Such gardens can be beautiful and have their value, but they are 'architecture' rather than 'horticulture'. True gardeners love growing things and living beauty, taking pleasure in the whole process of ever-changing life. Plants have their own beauty—they are life. This is what gardening is about and is also what growing roses is about.

2

ENGLISH
CLIMBERS
IN THE
GARDEN

THE ROSE PROVIDES some of the most useful and beautiful of all climbing plants and in recent years we have developed a fairly wide selection of English Roses that can be used as climbers. With their beautiful Old Rose flowers, their fragrance and their long period of flowering, they are a very useful addition to the Climbing Roses in general. They can be depended upon to grow to 2.4–3m/8–10ft in height and a little less across, which is about as tall as most gardeners require, and they will often grow much larger. If a climber grows taller than 3m/10ft, it may be difficult to manage—not everyone likes to prune and tie up a rose at dizzy heights—and, of course, individual blooms are more easily viewed at eye level. Nonetheless, we do have one or two varieties that, with generous treatment, will reach the roof and we are developing more tall climbers.

All English Climbers have typical English 'Old Rose' flowers which, in most varieties, hang gracefully on the branch, looking down upon us to excellent effect. Being not too ambitious in terms of height, they have the ability to flower with greater regularity and abundance than most Climbing Roses. With suitable treatment, they will usually provide three crops a year, unlike so many climbers that are not truly repeat-flowering. English Roses are seldom more beautiful than when they are grown as climbers. Most of them are remarkably free from disease.

As well as the English Climbers there are a number of English Roses that are usually grown as shrubs, which can also be grown as climbers. These we describe as Shrub/Climbers. 'Graham Thomas' is an example of such a rose. In fact, it is hard to know where to place it—as a shrub or as a climber. Whenever a shrub is capable of climbing, I mention this fact in the descriptions in Part II of this book. You may expect a Shrub/Climber to grow to 2.1–2.4m/7–8ft on a wall although, under the right conditions, they may go a great deal further. Sometimes they need a little encouragement if they are to climb well. They often produce a whole cluster of shoots at the base. Select a few of the longer ones and cut back the surplus shorter shoots, to encourage taller growth. Indeed, there is much scope for art in the way we prune and train roses—encouraging a branch here and cutting back another there—to provide the most pleasing effect. Further information on cultivation and training will be found in Chapter 6.

There is endless scope for ingenuity in using English Climbers and

English Roses in the Garden

Shrub/Climbers. If you wish to soften an otherwise hard structure in the garden, or if you are trying to create a rampant 'growthy' effect, then there is no better subject.

Walls and Buildings

Of the various structures upon which English Climbers can be grown, the most important are the walls of the house and buildings in general. Roses always look best when closest to where we live— and we do not get much closer than this. Here the formality of architecture goes well with the rampant growth and softly coloured flowers of an English Rose. A wall also has the effect of drawing roses to a greater height than we would expect elsewhere. This can be particularly advantageous in the case of the shorter growing Shrub/Climbers.

English Roses grow equally well on south-, east- or west-facing walls—and even on north walls, rather surprisingly, they do remarkably well. In spite of this, it is important that they should never be overhung by trees.

I will not suggest particular varieties for growing on walls, as nearly all English Climbers do well in this position.

Pillars

Pillars can make useful features and focal points in a garden and there is no more suitable rose with which to clothe them than English Climbers that flower with great regularity and grow to manageable heights. Pillars may be constructed out of timber, brick or stone. Alternatively you can purchase structures, which may be of timber or metal bars interlaced to give a lattice-like effect. They are usually some 45–60cm/18in–2ft across and approximately 2.4m/8ft high.

If you wish to bring height to a border, you might consider sinking posts along its length and growing English Climbers on these. From ground level the posts should be about 2.4m/8ft high. It is a good idea to use two plants on each pillar to ensure that it is covered with ample growth. The roses will sometimes cascade down to give a pleasing effect.

There are many other positions in which pillars and posts can be very useful. These are very much the same as I recommend for Standard Roses on page 250–1.

English Roses in the Garden

Trelliswork, Fences and Rustic Work

English Roses growing on trelliswork, rustic work or fences are often used to divide one area of the garden from another, or to provide privacy. A whole range of English Climbers can be used on structures of this kind, but it may be better to use taller-growing varieties. One advantage of trelliswork is that little tying is necessary. We can simply weave the growing shoots between the bars and these will do the work for us.

It is but a small move from a division made of trelliswork to a simple fence. A barred fence can look a little bleak, but is an ideal structure for Climbing Roses. English Climbers, being not too tall, are perfect for this use, and here the shorter English Shrub/Climbers come into their own, although the taller varieties can also be used. A certain amount of latitude may be allowed when pruning and tying back English Climbers on a fence. They may be trained neatly along the bars or encouraged to grow freely so as to tumble down over the fence in happy abandon, providing a rampant, billowing display.

In the case of rustic work, only the strongest English Climbers are suitable and it may be better to use other strong-growing Climbers and Ramblers if the structure is very tall.

OPPOSITE, TOP 'Teasing Georgia' forms an excellent shrub but is equally good as a Climbing Rose — particularly on a wall.

OPPOSITE, BOTTOM 'Tess of the d'Urbervilles'. Good red climbers are rather scarce at present among English Roses, but this variety is a good short climber.

BELOW 'Graham Thomas', one of the most popular English Shrub Roses, is even better as a climber. Note how nicely the blooms are held.

English Climbers in the Garden

Arches and Pergolas

There are occasions when an arch makes a very useful contribution to the garden. They are best placed so as not to appear too contrived, perhaps between one part of the garden and another. Arches are effective placed at intervals along a path where it is pleasing to walk beneath them and see them hung with fragrant bloom.

Arches require rather taller climbers than we might expect. Remember that they have to climb not only up the arch, but also across—and this is rather a long distance. Useful varieties include 'A Shropshire Lad', 'Malvern Hills' and 'The Generous Gardener'.

A well-clothed pergola can be a magnificent feature in the garden. Unfortunately, most English Roses that we have at present are not really suitable for growing on them. The effort of growing a great distance and flowering throughout the summer is rather too much for any rose of this kind. Only Ramblers really have sufficient strength, and, of course, these flower just once in a season.

Scrambling Roses

There can be few finer sights in the garden than a vigorous Rambler Rose, such as *Rosa filipes* 'Kiftsgate' or 'Paul's Himalayan Musk', that has been planted against a tree and encouraged to scramble up it and send down long cascading branches covered with bloom. English Climbers can fulfil a similar if more humble role planted at the foot of shrubs and permitted to grow over them. Many shrubs flower early in the year and can be a little boring later in the season. If you encourage roses to scramble through and over the shrubs, you will have flowers throughout the remainder of the summer. The roses should not smother the shrubs and the shrubs should not out-grow the roses, so it is vital that you select partners of similar vigour.

A similar effect can be made by encouraging English Roses to scramble over hedges. You would probably not wish to do this on a large scale, but there are situations where it can provide a pleasing effect. It is usually not necessary to control the roses to any great extent. It is better to let them have their own way, just occasionally removing a shoot here and there and training another along in a suitable direction.

ABOVE 'St. Swithun' is an excellent climber that will achieve 2.4m/8ft —more in a warm climate.

OPPOSITE 'Malvern Hills' has all the attributes of a Rambler Rose but repeat-flowers very well.

3

ROSE
GARDENS

FEW PEOPLE can afford the luxury of a garden devoted entirely to English Roses. They simply do not have the space, or perhaps even the time to maintain it. But for those who have, it can be a source of endless pleasure and interest. English Roses are so varied, in both flower and growth, that it is quite possible to devote a garden to them alone and we can be sure that it will nearly always be in bloom throughout the summer—although where the garden is large, we might consider mixing them with Old Roses, species roses, Rambler Roses and so on.

Even in a rose garden made up of various kinds of Shrub Roses, old and new, there is a case to be made for building this around English Roses. particularly if your taste tends towards the Old Roses. As I have said, most Old Roses are early-summer-flowering only and those that do repeat are often poor growers and subject to disease, as well as being limited in colour range.

Traditionally a rose garden implies square or rectangular beds arranged in some kind of geometric pattern, usually with one variety of rose in each bed. Such a design can be pleasing, but is really better suited to Hybrid Teas and Floribundas which are, in fact, bedding roses. With the rediscovery of the Old Roses and the arrival of the English Roses, as well as the introduction of a number of so-called Modern Shrub Roses, more imaginative schemes became possible and, indeed, necessary.

RIGHT A view across our small rose garden in Wales to the fields beyond.

OPPOSITE The concentric borders of the Victorian Garden at Albrighton. This picture is taken from a high angle. As you walk in the garden, there is a more intimate and less of a 'bedded out' effect.

English Roses in the Garden

The English Rose requires a very different kind of garden from that to which we have become accustomed. First I want to suggest broad approaches to the design of rose gardens and then to explore ideas based on my own experience of creating a small garden at our house in Wales and on the group of gardens we have built. Of course, other people will have other ideas and there is plenty of scope for gardeners to create something new and different.

The Principles of Design

The English Rose is not a standardised product: it is a shrub that varies widely in flower, growth, foliage and stature. We therefore have to think in terms not of neat, flat rose beds, but of a garden of interlacing borders containing roses of all sizes, so that we have a profusion of flowers with growth of varying height. An English Rose may be bushy and twiggy, or it may send up long, sweeping stems that arch gracefully. It may be tall or it may be short; it may be wide in growth or it may be upright. Such roses fulfil many different roles in the garden and require a very different kind of design.

Although it may not consist of formal beds, a garden of English Roses, or indeed of any other shrub roses, should usually be formal in character and yet make an intimate effect. I suggest a layout of interlacing borders. These may be of any width, but borders 1.8–2.4m/6–8ft wide are most suitable since the English Rose is an intimate flower and requires an intimate setting to show off its beauty to best effect. We wish to enjoy our roses as individual blooms as well as viewing them in the mass. One of the great pleasures of English Roses is their many and various fragrances. As I have said, most rose fragrances do not carry well upon the air and so have to be savoured at close quarters. Also, their fragrances are so varied that we like to enjoy them individually. All this requires a certain intimacy which, in turn, means relatively narrow borders.

English Roses, being robust, rather unruly or even sprawling plants, require a setting with some sort of discipline if your rose garden is to have form. I suggest that borders, and rose beds, should be edged with some kind of plant or 'hard' material within which the roses can grow with rampant abandon. The plants may be of box, yew, lavender or some other subject that lends itself to clipping. We find box and yew ideal, giving just the neat effect we require. Both yew and box have the advantage that they can be clipped to whatever height you like and will thrive on such treatment and keep their shape admirably. Alternatively, you can use brick, stone or some other hard edging material. This should be sufficiently substantial to make a firm statement. It is the contrast of severe form and natural growth that provides so pleasing an effect.

Whether your rose garden be small or large, it is often desirable to enclose it in some way, to create the intimate setting that best suits English Roses. Surrounding walls have the advantage that you can grow climbing roses upon them, but they are expensive to build. Trellis panels or rustic work may also be used and they, too, provide supports for climbing roses. A rose garden may also be enclosed by a hedge. Again, yew is probably the most suitable, as it can be closely clipped to give a neat outline to the garden. It also provides an excellent dark background which shows off the roses to perfection. However, you need to take precautions against the roots of any hedging plant invading the space in which the roses are growing, as these will impoverish the soil (see Growing English Roses, page 297).

BELOW Our garden in Wales shows how it is possible to maintain a sizeable collection of roses in quite a small area.

Small Rose Gardens

While many gardens are so small as to render a rose garden impossible without taking up the whole area, there are even more that have a large enough space—either at the front of the house or at the back—to make a small garden of English Roses of gem-like quality. If the back garden is of sufficient length, it may be divided in two so that you walk through a more general garden into the rose garden—or perhaps vice versa.

We have put our small rose garden in Wales at the front of the house. The

area is approximately 9m/30ft deep and 15m/50ft across, divided in two by a path leading to the front door. There is a small statue at the centre of each half of the garden, surrounded by a round bed. This bed is, itself, surrounded by a path and then borders, leaving spaces to enter at the centre at two points, so that in the middle you are surrounded by roses. The whole garden is edged by a low brick wall, over which scramble English Climbing Roses. Beyond the walls are the fields, which set off the garden to good effect. In a built-up area you might choose to enclose the garden with trellis, a hedge or a rather taller wall. The paths of our garden are of brick—red and black, arranged in patterns—and the borders are edged with low box hedges. The result is quite different from a garden bedded out with Hybrid Teas: the English Roses, with their shrubby growth, create a billowing effect which is far removed from a conventional rose garden.

A rose garden such as ours can be adapted to fit into all but a very small space. It could also form a pleasing feature in a much larger garden. If the roses are planted in groups of three of each variety, there will be space for quite a large collection of English Roses. There will always be some roses in flower throughout the summer—and for at least two periods of the year the garden will be flooded with bloom. For spring colour you might underplant the roses with low-growing bulbs like snowdrops, crocuses, miniature daffodils and so on, although this would entail using organic fertiliser rather than a mulch that would smother them. The work involved in such a garden is not excessive and it will be a source of pleasure for many years to come

A Garden of Double Borders

The simplest form of rose garden—but nonetheless, a very good one—consists of two borders placed opposite each other with a path between. This is particularly suitable for long narrow areas, like those behind many suburban houses, each of which has its own strip of land, but it could equally well grace any garden. The vista from either end can be of great beauty, with roses overflowing on to the path from both sides. The path between the borders may be quite wide—perhaps the same width as the borders—or it may be narrow, so that we almost have to wind our way between abundant roses; or it might be anything between these two extremes. The path may be a lawn, or it may be of stone, brick or gravel. I suggest using hard materials for narrow paths, where grass is likely to become worn with over-use. Where the path is wider, there is no such problem and grass is ideal.

The borders may be backed by hedges, fencing, trellis and so on, walls of brick or stone being the ultimate choice. Both trelliswork and walls have the advantage that they provide a home for English Climbing Roses. The background may be turned inwards at either end, leaving just a narrow entrance,

to create an enclosed rose garden. Alternatively, of course, you may decide to dispense with the background altogether, so that the borders are long island beds which can be viewed from all angles and from across the garden. A finishing touch might be a statue or some other feature at the end of the walk, so that the garden leads to a focal point. Or, perhaps, there might be a gate of wood or wrought iron at each end. Your choice will depend, to some extent, on how the rose garden fits into the garden as a whole. If a rose walk of this kind can lead to another part of the garden, so much the better: this will give it a certain logic.

The Gardens at Albrighton

With more space available, many different schemes for rose gardens become possible. It is not necessary or desirable to lay down rules as to exactly what form these should take, but it might be helpful if I describe, with the help of

ABOVE The view from half-way along the Long Garden at Albrighton, showing a mixture of Old and English Roses. The three figures at the end, sculptured by Pat Austin, form an excellent focal point.

photographs and a plan, the various gardens we have made at our Nursery in Albrighton. I am not suggesting that they should be copied exactly, but they may inspire gardeners in the creation of their own gardens, even if they are working on a smaller scale.

Altogether we have four gardens covering 0.8 hectares/2 acres, plus a small garden of species roses. Each is a complete garden, surrounded by a conifer hedge; all but the garden of species roses have a formal layout—although, English Roses being what they are, the overall effect is anything but formal.

The Long Garden

This is the largest in our group of gardens, being about 85m/290ft long and 25m/85ft wide. It is not so much a garden of double borders as three gardens of double borders, each pair running alongside the other. At the end of the central walk a carving in the form of three girls dancing provides a focal point. The borders to either side of the central walk are 1.8m/6ft wide. On each side of these run two more double borders. Paths run cross-ways at intervals, linking the three sets of borders and providing further views.

The Long Garden's central path is of brick, while those between the outside pairs of borders are of grass. All these paths might have been of gravel or other hard landscaping materials—stone being particularly pleasing. We chose to edge our borders with small yew hedges, clipped to about 30cm/1ft in height. It is unusual to plant yew for low hedging, but as it is such an excellent plant for clipping into all manner of shapes and sizes—as, for example, when it is used for topiary—we thought it might respond well to being cut very short. It has turned out to be very successful as an edging plant. It has a tendency to expand over the years, but I am quite sure, from all my

RIGHT Rustic arches and a mixture of English and other roses in one of the side-walks in the Long Garden.

English Roses in the Garden

THE GARDENS AT ALBRIGHTON

Species roses

THE VICTORIAN GARDEN, made up exclusively of English Roses, consists of three circular borders which become progressively narrower towards the centre.

THE RENAISSANCE GARDEN, planted purely with English Roses, forms the centrepiece of the group of gardens, with paths running through four sets of double borders and a canal at the centre running up to a loggia.

The loggia

The dolphin fountain

THE LION GARDEN is a traditional formal rose garden with Hybrid Teas, Floribundas and English Roses in beds.
.

The lion sculpture

At the end of the central walk a carving in the form of three girls dancing provides a focal point.

THE LONG GARDEN consists of three sets of double borders containing a mixture of English Roses, Old Roses and Modern Shrub Roses, backed by Climbers and Ramblers.

My route through the gardens as described in the text.

experience with yew, that it could be cut back hard, if necessary, and still come back as green as ever.

The three double borders are divided by Climbing and Rambling Roses on rustic work. Beneath the cross-timbers and between the uprights there is a yew hedge of about 1m/3½ft in height. The result is a series of 'windows' through which we can catch glimpses of the other parts of the garden. Occasionally, across the whole, we have further rustic work which serves to break up the garden a little and add further interest. This also gives us the opportunity to grow Climbing Roses over arches where the rustic work crosses the paths. This garden never lacks interest and variation and there are plenty of intimate corners in which to place groups of roses.

The Long Garden, as a whole, plays the role of the hall of a house. From it run paths to other gardens, each designed to show a different way of displaying English Roses.

The Victorian Garden

If we walk through the Long Garden and turn to the left at the end, there is a view into the Victorian Garden. This consists of three concentric borders, the outer one being the broadest and the inner two becoming progressively narrower towards the centre. Each of these borders is crossed at six points

by paths, which run to the centre like the spokes of a wheel. At the 'hub' there is a carved stone figure of a woman holding a bouquet of roses. Looking out from the centre, each path offers a view of further carvings at the perimeter of the garden, four depicting the seasons of the year and one of a mythical figure known as the Green Man, the sixth view being back into the Long Garden. Each of the outer carvings in the Victorian Garden is covered by an arch or arbour bearing English Climbing Roses. The garden is surrounded by a conifer hedge. The whole garden offers a great variety of views from many angles and the roses are nearly all within touching distance—and can be viewed with ease. The garden itself is not dissimilar in principle to that which I have described as a traditional rose garden; that is to say, a typical modern rose garden of rectangular beds. However, the overall effect is very different, largely due to the fact that the English Roses are of varying height, character and continuity.

English Roses in the Garden

The Renaissance Garden

If we return to the Long Garden and turn right—retracing our steps—and again turn to the right, we have a view of the garden that we have, perhaps a little pretentiously, called the Renaissance Garden. This forms the centre-piece for the whole group of gardens at Albrighton. Like the Long Garden, it is divided into three sections. As we enter, we face a long, canal-like pool, which is about 1.5m/5ft in width, with broad stone edging and English Roses at either side, their growth and flowers tumbling on to the stone and, occasionally, showing their reflections in the water. At the far end of the pool there is a sculpture of a dolphin spouting water. Beyond is a stone loggia—a temple-like summer house—from which to enjoy the vista down the pool and other views of the garden.

Behind the roses along each side of the pool are two wide grass paths with roses at either side growing in narrow borders with sinuous, snake-like inner edges, defined by low box hedging. The paths offer two further vistas from the loggia that end with a statue. In each loop of the snake is one variety of English Rose. This shows them off to maximum effect and

The canal in the
Renaissance Garden is longer than
this photograph suggests. The classic
arches of the loggia seem to suit
the garden to perfection.

provides an ideal opportunity for visitors to choose roses for their own gardens. We have used only the smaller-growing varieties and all these roses are closely pruned so that they fit into the confined space.

The snake borders are backed by low yew hedges and, from the wide grass paths in the Renaissance Garden, we can catch glimpses of the roses on the other side. As we turn left or right at the end of the paths and explore further, we come across a quite different way of growing English Roses. Here we have massed the roses in large swathes so that, with a grass path curving through them, we get the effect of walking through waving corn. English Roses, with their natural growth, are particularly suitable for this kind of massed planting. Its overall effect is dramatic, while offering intimate contact with the roses. It is surprising how well English Roses grow under these conditions. It seems that they enjoy the competition, each with the other, and the competition seems to result in altogether taller plants. The whole area is a riot of colour. This kind of massed planting has been used very effectively in the Rose Gardens at Alnwick Castle in Northumberland.

The Lion Garden

Earlier I suggested that a garden of formal rose beds is not the most suitable for English Roses, since it does not give them the opportunity of showing off the beauty of their growth. Nonetheless, if we resume our walk back down the Long Garden, turning right we come to a formal garden containing both English and Modern Roses—and it does have its charms. Some people prefer a rose garden to be formal and perhaps find it more suitable to fit into their garden as a whole—or they may already have such a garden and wish to retain it.

Although English Roses are shrubby in nature, some varieties are quite short in growth. These varieties also flower with remarkable continuity, which is an important quality in a bedding rose. In fact, we find that some of them flower with even greater freedom than the Hybrid Teas and Floribundas. If I have a complaint about English Roses when used for bedding, it is that most of the suitable varieties tend to be rather muted in colour and I have the feeling that bedding requires bright colours. Nonetheless, our formal Lion Garden of mixed shades of English Roses gives a pleasing, dappled effect. With English Roses there is no need to give gardeners the usual advice to restrict each bed to roses of one variety so as to avoid colour clashes. However, if we wish to place English Roses alongside Hybrid Teas and Floribundas, it is usually best to choose these in similarly gentle shades or to keep the brightest and most intense colours apart from the English Roses—although the occasional bright yellow, scarlet or flame shade will serve as a contrast which can help to lift the garden as a whole.

Our formal rose garden is situated between two borders of shrubby English Roses mixed with tall herbaceous plants—tall so that the plantings balance with the low rose beds.

Great Rose Gardens

It is very satisfying for people like myself, who have spent much of their life with roses, to note the number of large public rose gardens that we can enjoy today. A few of these, such as La Roseraie de l'Haÿ-les-Roses near Paris, Gardens of the Rose at St Albans, England, and Sangerhausen in Germany, seem always to have been there, but in recent years great rose gardens have grown up in many countries around the world. Not only this; many great private houses have superb rose gardens which are open to the public. That there are such gardens is a great compliment to the rose and provides us with ample proof of its beauty and garden value. It also confirms the very special place the rose holds in our hearts—if such proof were needed.

The English Roses can play a very useful part in a great rose garden,

however large. Indeed, a rose garden of any size may, if we wish, easily be made up of English Roses alone and guarantees a great deal of variety and a continuous display of rose beauty throughout the summer. In a garden on a very large scale, however, using English Roses alone would involve either a great deal of repetition or vast swathes of each variety. The family of the rose is so large and so various that I think it would be a shame to confine it to one group of roses of any kind. It would be equally wrong to make a great garden with Hybrid Teas and Floribundas alone.

If we consider the family of the rose, we cannot but be struck by its diversity, starting with the daintiness of the rose species, moving on to the charm and fragrance of the Old Roses and the English Roses, and on again to a whole range of Modern Shrubs—to say nothing of the Hybrid Teas and Floribundas. To these we must add the many Climbing and Rambling Roses. If, for the time being, we leave out the Hybrid Teas and Floribundas (for these really need a part of such a garden to themselves), we find there is a problem. This is the fact that the majority of the most beautiful Old Roses, Shrub Roses and species rose, do not repeat-flower. It is true that some Old Roses, such as the Chinas, the Bourbons and the Hybrid Perpetuals, do repeat in varying degrees, but they generally do not repeat well and have a limited colour range. It is for these reasons that English Roses can make a valuable contribution to a large mixed collection. Planted with Old Roses and Shrub Roses, they bring colour and interest when other groups have ceased to flower. They also greatly increase the colour range.

The great problem with a very large rose garden, containing many different roses, is the simple fact that it is so large and tends to be impersonal, which is not the ideal setting for an intimate flower. I suggest, therefore, that large areas be divided into a number of rooms separated by hedges or walls, as we have done at Albrighton. This offers the intriguing experience of walking from one garden into the next, each of which can be of a different nature. These rooms might include all the kinds and styles of garden described in this chapter—and I have no doubt that many other designs are possible.

In addition to providing us with the opportunity to create many different styles of garden, the great rose garden enables us to display all the many different classes and groups of roses that have been passed down to us over the ages. With the exception of the Hybrid Teas and Floribundas, all these roses mix together in harmony. I would not recommend putting the various individual groups of roses into gardens of their own, or even arranging them in the order of their history, for interesting though this would be, it would mean that some of the gardens would be devoid of flowers for the greater part of the summer.

OPPOSITE Evening in The Lion Garden, with 'Sophy's Rose' in the foreground, showing that English Roses can be very satisfactory when hard pruned and grown for bedding. The sculpture of a lion is by Pat Austin.

ENGLISH
ROSES
IN THE
HOUSE

ENGLISH ROSES are not only good garden plants; they are also among the best flowers for cutting and arrangement in the house. They are less rigid in growth than the Hybrid Teas and this gives them a certain grace when arranged in bowls. Their open flowers come in a whole range of different formations: single, semi-double, rosettes and cups, as well as spray-flowered. This in turn makes possible a greater variety of arrangements with far less of the sameness that we have with the Hybrid Tea Roses. The fact that the flowers of English Roses are softer in colour and texture than the Hybrid Teas enables them to mix more easily with other flowers as well as with one another. The play of light between the petals provides perhaps even better effects in the house than it does in the garden. If we add to this the many and varied fragrances of English Roses, we have what amounts to a very good cut flower. Indeed, one leading London flower arranger, Shane Connolly, once said to me that if he could obtain English Roses more easily, he would use nothing else.

Cutting for Arrangement

The best time to cut roses is in the morning before the sun has reached full strength. At this time the stems are full of moisture and for this reason the flowers are likely to last longer. The stage at which the blooms are cut is important: the flowers should neither be too open nor too closed, but about half-way open—although this depends on the variety. If the blooms are too closed when cut, they will probably never open properly and will never make mature flowers; if they are picked when they are too far developed, the petals will soon fall, although, again, this depends on the variety. There is a happy medium which will become obvious with a little experience.

When cutting roses, it is a good idea to have a bucket or some other container nearby so that the stems can be immediately placed deep into water. Experiments have proved that the stems begin to callus over almost as soon as they are cut, thus making it difficult or even impossible for the rose to take up water. Some people cut the stems a second time *under* water to avoid any chance of this happening. If you choose stems that are thicker, they will hold water better and the flower will last longer. Unfortunately, blooms with thick stems tend to be more rigid—and therefore often less

OPPOSITE A classic bowl of 'Eglantyne' with forget-me-nots.

English Roses in the Garden

graceful—and so a compromise sometimes has to be made. Short-stemmed flowers also tend to last longer since the moisture has less distance to run— but, of course, such stems are of limited use. We require stems of all lengths.

It is a good idea to use cut-flower care solution in the water to extend the life of the rose. This slows down the decay of the stems which, once it occurs, restricts the flow of moisture to the flower. Even then, the ends of the stems will inevitably begin to decay as the flower ages, again depriving the bloom of moisture. For this reason, it is a good idea to cut the ends of the stems again after a few days in water.

Arranging English Roses

It is helpful to have some idea of what you intend to do with your roses before you start cutting, so that you can pick accordingly. The range of possibilities is extensive, the very simplest arrangement being the posy bowl or perhaps just one bloom in a bowl. For those of us who take a special interest in roses, this is an excellent way to appreciate the individual flower.

English Roses in the Garden

I often take a bloom here and a bloom there from our trial grounds and place each in a small vase so that I can get to know them more completely. If you have a number of such bowls that look well together—perhaps of varying size and shape—you can place them on a table or elsewhere and shift them around to form an attractive group.

Another simple idea is to cut the blooms with very short stems and float them on a shallow dish or bowl of water, to provide a water-lily-like effect. The flowers will require a little support if they are not to become waterlogged. Foliage placed on the water can be useful here and at the same time makes an attractive background. Alternatively, you can use layers of slightly crinkled wire netting and poke the short stems through this.

For larger arrangements, a great variety of effects is possible. The receptacle you choose will have a considerable impact on the arrangement. It is easier to arrange roses in a tall, narrow vase than in a shallow bowl. If you pick a bunch of roses that seem to go happily together and simply 'dump' them in a vase, you will give yourself the beginnings of an arrangement. The blooms can then be shuffled around, but it is usually best to let the flowers have their way as far as possible and try to enhance the arrangement with other blooms—one added here and another moved there—until a pleasing picture emerges.

Arranging a broad bowl of roses requires more skill but offers greater scope and a low bowl is, of course, essential for the dinner table, if we are not to blot out the person sitting opposite. Some sort of holder will be useful to keep the blooms in place. These are available from florists and come in a number of different materials and designs, most of which are effective. It is a good idea to have various deigns to hand. Alternatively, you can use crumpled wire netting, formed into a mound, and push the stems in between the wire. The stems must be sufficiently deep in the water for the blooms to survive. It is essential to avoid a too-upright appearance—try to get breadth into your arrangement. Allow the roses to spread gracefully and, to some extent, nod upon their

English Roses were at first largely in the softer colours which are, perhaps, most typical of them. In later years we moved on to breeding warmer colours as well. Here we see an exciting arrangement of such colours, including 'Pat Austin' (*centre*) and 'Crown Princess Margareta', with florists' flowers.

stems. Most English Roses have a natural grace try to make the most of this.

There is currently a fashion for massing blooms in a bowl so that the roses are packed together with little or no space or foliage between them. The effect can be pleasing, but the full beauty of individual blooms is lost in the mass. For the dinner table it is an easy solution, but on the whole I do not rate such an arrangement very highly.

I cannot pretend that roses are the easiest of flowers to arrange. They tend to have a mind of their own and to look in the direction that they wish rather than you would wish. While most of the blooms will slot in as you desire, it might take a little time and experimentation to find the roses with the correct stem shape to complete the picture. An arrangement should not be too stiff and formal, nor should the blooms be bunched together in one area and too much space left in another. The roses should be encouraged to show off their natural beauty. The lighter-weight flowers can be used to balance and join together the heavier, more opulent blooms into one picture.

The careful use of the various groups of English Roses—the Old Rose Hybrids, the Leander Group, the English Musks and the Alba Hybrids—can add interest and variety to an arrangement. As in the border, spray-flowered varieties, such as 'Blythe Spirit', 'Scarborough Fair' and 'Francine Austin', as well as semi-double roses like 'Marinette', 'The Herbalist', 'Windflower' and 'Windrush', can be helpful as a balance to the typical heavier English Rose.

Of course, it is not necessary to confine an arrangement to English Roses alone. The Old Roses, in season, will mix happily with English Roses, being of a very similar nature. There are also many other roses—such as the Rambler Roses, Ground Cover Roses and some Hybrid Musks—which can be used to good effect with English Roses. These should not be used too generously, but a spray here and there, between the English Roses, can add much to the beauty of an arrangement. The Hybrid Teas, however—beautiful though they are and much as I would hate to be without them—do not mix easily with English Roses. Their beauty is of a different kind. They are

ABOVE A bowl of white roses can be pleasing. Here are 'Winchester Cathedral' and 'Francine Austin' with sprays of blackberries.

OPPOSITE English Roses can be particularly attractive when the light shines through their petals from behind. These are 'Golden Celebration' and 'Blythe Spirit', with aconite and lavender.

English Roses in the Garden

English Roses in the Garden

hard-edged flowers and usually of intense colouring, which means they are probably best kept for a bowl of their own, although no doubt there are plenty of exceptions. As with so much in roses and gardening, we find out only by experimenting.

English Roses are particularly useful in a mixed arrangement of herbaceous and shrubby flowers. I will not attempt to make suggestions here, as the possibilities are endless. Even the inclusion of two or three English Roses can greatly enhance a mixed bowl and some of the larger-flowered varieties are particularly effective for this purpose.

The question of colour combinations is very much the same as it is in the garden. One advantage, which we seldom have when planting a border, is that as we pick our roses we can place one against the other to see how they go together. Very broadly speaking, as in the garden, it is a question of warm colours and cool colours. The warm colours are those that suggest flame and fire—the bright reds, oranges and yellows. Cool colours include

ABOVE A beautiful arrangement of 'Teasing Georgia' and 'Mary 'Magdalene'.

OPPOSITE
TOP 'William Shakespeare 2000' and 'Charles Rennie Mackintosh' with Rosebay Willow Herb and Marjoram.
BELOW The new English Cut Flower Roses —'Olivia Austin' (pink); 'Juliet' (apricot); 'Oberon' (purple) and 'Portia' (white).

pinks, mauves and violets. If we keep the cool colours together, we achieve harmony, and the same is true for warm colours. Occasionally you may wish to create a contrast, by, say, adding a yellow flower to a bowl of cool-coloured roses, but this should done with care. If you place blooms of different English Roses together on a table, you will be surprised how one colour can help another, so that both are enhanced. Above all, the important thing is to find what pleases *you*—this is really the only way. On the whole, English Roses harmonise easily and it is difficult to go too far wrong.

I cannot close this chapter without some mention of the new English Cut-Flower Roses, which are available from some florists. These roses are described, and two are illustrated, on pages 236 to 237. Those of us who grow English Roses in our gardens require such roses only in winter. It is good to enjoy a glimpse of summer in the house during the darker months; at such times roses are even more welcome than usual.

Commercial cut-flower roses have always seemed to me to be rather stiff and lifeless; one might say that they are almost manufactured in appearance and somewhat lacking in variety, except for colour. We hope that over the course of time it will be possible to bring some of the natural beauty of the English garden roses to cut flower roses.

THE
FUTURE
OF
ENGLISH
ROSES

SINCE I FIRST began the development of what we now call English Roses, our breeding programme has steadily expanded year by year. It now involves the annual hybridisation of some 150,000 flowers. Not all cross-fertilisations succeed in producing seed but, from those that do, we grow about 250,000 seedlings each year.

Our seedlings are grown on benches in greenhouses and will flower when they are still quite small, within about four months. From these we select the ones that show some promise, about 8,000 garden roses and a similar number intended for cut flowers. These will then be budded on to stocks in our trial grounds and will form bushes in the following year. During the next three years, a careful assessment is made of the plants for all the characteristics we aim to produce in our roses. The most promising seedlings are then propagated in larger numbers and are observed for a further three years. All this time we are looking for the particular beauty and character of flower we wish to have in English Roses. We also observe growth: is a particular rose likely to form a pleasing shrub? At the same time we consider the leaves—attractive foliage has so much to contribute to the beauty of the flower and the shrub as a whole. Next we look for fragrance not only a strong fragrance, but a beautiful one.

As we estimate a rose's aesthetic merits, there is also an array of practical points to consider. Has a rose got sufficient vigour to grow well, even for the less experienced gardener and under less-than-ideal conditions? We note its ability to flower repeatedly and freely throughout the summer. Last, but by no means least, we study the rose for its resistance to disease. This we consider vital and we look forward to the day when most of our roses are largely disease-resistant and spraying becomes unnecessary.

All the above factors are noted first on hand-held computers and later transferred to a central computer in the office. Gradually, over the seasons and the years, we build up a picture of each plant and its abilities. When we are satisfied with a particular seedling, this is propagated in quantity in our rose fields and will be sold the following year. The whole process, from hybridisation to the day a rose is offered to the gardener, takes a minimum of eight years. The result is about six new roses each year. All the other seedlings—many of them of considerable beauty, though lacking in some vital quality—are consigned to the bonfire.

The Process of Breeding Roses
BELOW **First year** (**1**) A Seed House.
(**2**) Collecting pollen. (**3**) Pollen ready for
hybridising. (**4**) Hybridising. (**5**) Rose hips.
OPPOSITE **Subsequent years** (**6**) Newly
germinated seedlings [second year].
(**7**) The first small flowers appear in the

Roses intended for cut flowers take a little less time, as they spend their whole life under glass and can thus be assessed more quickly. Our requirements for beauty of flower and fragrance are the same, but we lay great emphasis on a rose's productivity and lasting power when picked.

Once a rose is considered to be of a sufficiently high standard, budding wood is sent to other nurserymen in all parts of the world. They will assess the roses for their ability to grow in climates other than the British Isles, and provide us with information to help us build up a more complete picture for future development.

The Future

Uppermost in our minds throughout the whole process of developing a new rose, at least as far as garden roses are concerned, is the beauty of the shrub and of its flowers, together with its fragrance. We do not believe that the rose has anywhere near reached its zenith in these respects. Always we are looking for better garden plants—ones which will associate well with other plants and fit happily into the general garden scene. With these factors in mind we decided to develop our new groups—the Old Rose Hybrids, the Leander Hybrids, the English Musk Roses and the Alba Hybrids, together with the English Climbers—each of which is of widely-differing character. Once these groups have settled down, which will take some years, we aim to breed each along its own unique line, so that it has a special beauty and a distinctive place in the garden.

Looking a little further into the future of English Roses, the possibilities seem to be infinite. The perfect rose is, fortunately, impossible. A bloom can always be chiselled out to be a little closer to a dream of beauty. This is in the nature of the rose. It has a wide variety of species with differing flowers and growth. Man has bestowed upon it its numerous petals, and these can be formed into all manner of shapes.

There is no great virtue in variation for its own sake. There are some forms, however, that are poorly represented among English Roses: for example single and semi-double flowers are somewhat lacking. These are capable of much greater beauty, as well as a greater variation. They depend very much on the shrub upon which they are produced for their elegance and charm. Add a few more petals and we get a flower that comes closer to the peony in its appearance. Flowers with quite a large number of petals, but still displaying their stamens, seem to me to be some of the the most beautiful of all. These barely exist and offer a very worthwhile field of development. A boss of golden stamens seems to light up the whole flower and give it a central focus that is entirely beneficial.

A further area of potential development is of the bud flowers of the

Hybrid Tea. I have spent much of my life drawing people away from this kind of flower, but I would be very sorry to lose it. If we look at a well-developed flower of the rose 'Madame Butterfly' held upon a climber, we cannot fail to appreciate its beauty. I believe that if we could develop nicely formed buds that open into beautiful flowers of the English Rose type—and if these flowers were elegantly held on a shapely shrub rather than on a squat bush—we would have a very beautiful rose. It would not, however, be an easy rose to develop, although we do have one or two varieties that fit this description already.

Growth and foliage also offer us great possibilities; not only for their health and vigour, but for their own beauty and the manner in which they set off the flower. In fact, our aim is to look at the plant as a whole—the way in which all the various factors combine to create a thing of beauty.

Having said all this, it is not so much for new forms that we are looking, as for the refinement of those we have already into something ever more beautiful.

The Way Forward

I would like to conclude by putting forward a few thoughts on the breeding of not only roses, but of garden plants in general.

There are few flowers that cannot be enhanced in some way for garden purposes, although in some cases the scope for improvement is only slight and no flower should be pushed further than its capability allows. Some increase in size and showiness is possible, but this must be balanced with the natural beauty of the plant.

Many garden flowers—like so much that is to do with the adornment of our lives—are subject to the vagaries of fashion. This is a shame but, in the case of certain of our more developed flowers, understandable. The flower breeder is inclined to become the victim of market forces. He or she takes a flower that is perhaps little more than a creature of the wild and develops it into something larger and more showy. This often enhances its beauty and makes it more suitable for growing as a garden plant. Soon, flushed with success and perhaps with the financial returns that go with it, the breeder goes on to make the flower even bigger and brighter and the plant more and more floriferous. Societies are formed, rules are laid down, committees meet—and soon we have a cult. But there comes a time when the breeder's efforts are counter-productive. You cannot provide more beautiful flowers by mathematical progression. All too often the flower becomes too large and too gaudy, and begins to lose its beauty and appeal. We see this in dahlias, chrysanthemums and the much-abused gladioli. We see it in the rhododendron, despite its aristocratic connections, and, to a lesser extent in

greenhouse [second year]. (**8**) The roses are budded in groups of eight—here they are being selected for the first time in summer [fourth/fifth year].

(**9**) The selected roses are budded in rows of two hundred, for final selection [sixth/eighth year].

ABOVE An exhibit of English Roses at Chelsea Flower Show, where they have won numerous Gold Medals over the years.

INSET Michael Marriott of David Austin Roses, who has done so much to make English Roses known to gardeners.

the magnificent peonies, irises and delphiniums, all of which are in constant danger of moving a stage too far. All these plants are magnificent in their different ways, but those who have the responsibility of maintaining them should be wary of spoiling them.

The rose is almost unique in its seemingly unending capacity for development. It is important that all who make it their business to create any new flowers should see their job more as an art and less of a science, useful though science may be on occasions. In this—if it is not too arrogant— I would like to think that the English Roses are, if only in a small way, pointing out the road ahead.

The people who have made the English Roses possible include, in one

English Roses in the Garden

way or another, all the management and staff of our Nurseries. Among them, Carl Bennett has been particularly important in recent years. Still in his early thirties, he has managed our Rose Breeding Department since 1996 and has a vast knowledge of the many different strains of English Roses that are always in the course of development. He is a devoted enthusiast and entirely dedicated to their future. He has a remarkable memory for everything that is going on—a most important skill in plant breeding—and, even more importantly, he has a genuine appreciation of what we are looking for in English Roses. He is backed by an excellent and improving staff of some twenty people.

Most importantly of all, I must mention the other half of our business, my son, David J.C. Austin. Since David joined us in 1989, he has become the driving force behind a large proportion of what we have done on our Nursery. Much the greater part of our success as a business has been due to his efforts in recent years. There has been, perhaps, too little mention of him in this book. His knowledge and ability have made a very important contribution to English Roses. He also has a good eye for a rose and, indeed, a deep love not only of roses, but of gardening in general. It would be hard to think of a better person to work with. You may be assured that the English Roses are safe in his hands.

ABOVE David Austin.
LEFT Taking a break at Chelsea Flower
Show with my wife Pat and son
(David J.C. Austin).

The Idea of the English Rose

6

GROWING
ENGLISH
ROSES

Many people reading this book will be well versed in the art of growing roses. This chapter is for those who are not, although experienced gardeners may find points that will be helpful. The advice I offer is something of a counsel of perfection. English Roses can be grown quite well with the application of a little common sense and it is not necessary to make what should be a pleasure into a burden. However, you cannot expect the best results if you simply plant them and forget them. Like many other highly developed flowers, they require a degree of cultivation if they are to give of their best and even a little extra attention will bring great improvements. This is part of the pleasure of growing roses, at least for those of us who enjoy gardening.

Growing English Roses is much the same as for any other roses. There are, however, important differences regarding planting, pruning, feeding and disease control.

Methods will, of course, vary somewhat in accordance with the part of the world in which you are growing roses. If your climate is very different from that of northern Europe, you need to consult a book on growing roses in your area.

Preparation for Planting

There are two ways to purchase roses —bare-rooted or in pots. Some people have the idea that a bare-rooted rose is in some way inferior to a pot rose. This is not true; the bare rooted rose is probably marginally superior, if planted at the right time.

THE TIME TO PLANT In temperate regions bare-rooted roses are best planted in late autumn (November in the UK), but they can be planted at any time up to late spring (the middle of April in the UK). When planted late they will flower a little less freely during the first year, although they will have caught up by the second year. Container roses—those that are sold in pots—can, of course, be planted at any time of the year if the soil around them is kept sufficiently moist.

WHERE TO PLANT English Roses are vigorous and therefore easy to grow and will thrive in a variety of positions and soils. Nonetheless, it is important to take the following points into consideration when deciding where to grow them. First, it is best to choose a site where the soil is deep and fertile. Rather surprisingly, English Roses can be grown in positions where they will enjoy only a few hours of sun throughout the day. However, avoid planting them where they will have too much competition from other plants and shrubs, particularly where these are strong and invasive. Like most roses today, they are expected to flower throughout the summer and this makes exceptional demands upon them. In a mixed border make sure that other plants growing around them are not too close and not too vigorous. If plants are allowed to invade their space, few roses will succeed. Roses should never be planted under trees, or be expected to compete with the roots of trees.

If you are digging up old or unwanted bushes, it is important to remember that roses will not thrive in a position in which other roses have previously been growing. It is impossible to over-stress this point. Even if the previous roses were growing well when they were dug up, the new roses are still unlikely to thrive. This is due to an infection in the soil, popularly known as 'replant disease'. Either choose a position where roses have not been grown recently, or dig out the soil to a depth and width of 50cm/18in and replace it with 'clean' soil from elsewhere.

PREPARING THE SOIL Having chosen the position in which to plant your roses, dig the soil to a depth of 30cm/1ft, add liberal quantities of humus and mix it in well. The humus may be well-rotted farmyard manure, garden compost or a proprietary compost from your garden centre. The key word is 'liberal'. It is particularly important to add this bulk to light, sandy soil and chalky soil. If the subsoil is compacted, break it up with a fork to improve drainage and to assist the rose in sending down its deep tap roots, but do not add humus at this depth. As regards the question of lime content of the soil, roses prefer a slightly acid soil with a pH of 6.5.

Planting English Roses

I regard the planting of English Roses in groups as being of first importance and have explained why on pages 242 to 244. I would like to emphasise this by summing up that advice. Being small shrubs which, like other roses, are budded on root stocks, English Roses are a little untidy—rather narrow at the base and broad and heavy at the top. Planted in groups of two, three or more plants of one variety, very close together —with no more than 60cm/2ft between the plants in a group—they will grow into what amounts to one fine, well-rounded shrub with three root systems, which will enable it to flower freely and continuously throughout the summer.

It is important to dig a hole large enough for the roots of your rose bushes to spread out without bunching and deep enough for the soil to be 75mm/3in above the point where the roots meet the green growth—that is, where the rose was budded. Work a little soil—which should be moist but not too wet—in between the roots, fill in the hole and then gently firm in with the feet. Planting in very damp soil should be avoided, otherwise it can set and become airless and the rose may remain semi-dormant in the first year. If necessary, purchase a planting compost from your garden centre and place this around the roots—or you might find some dry soil in your garden close to a wall which you can use for the same purpose.

Roses near Hedges

I have described the merits of surrounding a rose garden with clipped hedges, and mentioned the cultural problems of yew in particular (see page 269). To prevent the roots from invading the area where the roses are growing, and robbing them of moisture and nutrients, you can sink 75cm/2½ft sheets of galvanised iron, vertically, into the ground close to the hedges. (Bend over the tops of the sheets to avoid leaving a sharp rim.)

Pruning

Pruning is not quite the difficult task it is sometimes made out to be. English Roses require rather different treatment from Hybrid Teas and Floribundas and are, perhaps, less exacting. Pruning is best done earlier than is usually recommended to give your roses ample time to get in two full crops of bloom. The timing varies according to the climate. In the UK and elsewhere with relatively mild winters, late December, January or February are the best times. In regions with cold winters pruning should be delayed until spring growth is just starting.

PRUNING IN THE FIRST YEAR Bare-rooted roses will usually have been sufficiently pruned before you receive them. If not, they can be cut down to 45cm/18in in height. No further pruning is required in that season. If you have purchased roses in pots, they will not need pruning until the first winter.

PRUNING IN SUBSEQUENT YEARS This will vary according to the variety of rose and what you require it to do. English Roses are very accommodating in that they can be pruned long to produce a tall shrub or short to produce a smaller shrub. This may sound like an obvious statement, but it would not be true for Hybrid Teas and Floribundas or, indeed, for most other roses.

LONG PRUNING The most usual way to grow an English Rose is as a shrub, generally of some 90cm–1.5m/3–5ft in height, according to variety, although some varieties will grow much larger. To achieve shrub-like growth, it is only necessary to cut your rose back by one third of its height. This is not a highly scientific operation—if you have many roses, you can even use a mechanical hedge cutter, although secateurs would be marginally better. Having done this, all you need do is thin the shrub out a little by removing some of the weak and twiggy growth and at the same time cutting out all ageing, dying and dead growth. By 'ageing' growth, I mean main branches that have almost ceased to produce robust flowering shoots. This is the important part of the pruning of English Roses: removing rather than cutting back growth. Any further pruning is largely done in order to give the shrub a satisfactory shape. Try to vary the length of your pruning according to the height of the variety. This will give a more natural and less flat appearance to the border as a whole.

Pruning English Roses is as much an art as it is a craft. It gives you the opportunity of shaping a variety—or, better still, a group of one variety—in such a way as to bring out its natural beauty. If you observe your roses over the first year or two of growth, it will soon become clear whether you are over- or under-pruning. The most important consideration is whether or not your rose is beginning to form an attractive shrub. When taller English Roses are under-pruned, their branches will sometimes flop over and look unsightly, but this can usually be overcome by rather harder pruning. At the same time it is important to be careful not to destroy the effect of a nicely arching shrub.

HARD PRUNING If you are growing English Roses in a formal rose bed, or in the front of a border or a confined space, you can prune them much harder, taking away some two thirds of their growth. While it is not possible to turn a very large shrub into a very small one, you can, if you wish, prune an English Rose that is not excessively tall very much as you would a Hybrid Tea. This will, of course, mean that the shrub will be less shapely, but in some positions this does not matter very much. In all other respects the procedure for hard pruning is the same as for long pruning.

DEAD-HEADING The removal of dying and dead flowers is important. If you do not dead-head, there will be a tendency for your rose to produce hips, and these take the strength out of the plant, which would otherwise be used to make further flowers; and with some roses, you may not even get a second crop at all. When dead-heading, it is a good idea to take away a little of the stem so that the new flower shoots will come from good, strong growth. At the same time, you can tidy the plant up a little.

REMOVING EXTRA-LONG BRANCHES Some varieties of English Roses occasionally send up tall, ungainly branches, particularly in countries that have long, warm summers where the vigour of English Roses can result in too much growth. This is because they were originally bred from a mixture of Old Roses and other roses, as well as short Climbing Roses, although this is now some

generations ago. Long shoots are not likely to be numerous but they can make the shrub look unbalanced and unsightly. All you have to do is to cut the branch back to just below the average height of the shrub.

This process can be carried further and the roses given what is known as a 'summer pruning'. It should still be only a light one.

SUCKERS Shoots that come from the root stock rather than from the budded rose should be removed as soon as they appear, otherwise they will drain the life from the rose. They should be cut out as closely as possible, even taking with them a little chip of the stock so they do not shoot again.

Maintenance

Roses need generous treatment since, being repeat-flowering, they have a lot of work to do. Plenty of humus and an occasional feed with fertilisers will be amply repaid. Watering, too, is advantageous.

MULCHING We are always told that the rose is a gross feeder, and this is correct. Though not essential, there is no doubt that mulching—placing a thin layer of *well-rotted* farmyard manure, garden compost or some form of proprietary mulching compost around your roses—is very beneficial. In fact, it can improve your roses out of all recognition. Mulching plays an important role in helping to keep the soil cool and moist throughout the summer and feeds your roses at the same time. The mulch should not be very thick; too great a depth tends to exclude air from the soil.

FEEDING While a mulch supplies many of the nutrients the rose requires, a sprinkling of rose fertiliser around the bushes corrects any soil deficiencies and helps you to get really good flowers. More importantly, a further dressing of fertiliser should be applied just as the first flush of flowers is almost complete. This will encourage

new growth for the next flush. As I have said, a great deal is expected of a rose in that it flowers throughout the summer and to do this it requires ample nourishment.

WATERING Unlike most wild roses, English Roses (and many other roses) are repeat-flowering. No plant can continue to flower without nourishment and all the feeding in the world will do no good unless there is moisture to make it available to the plant. For a truly superb display of bloom throughout the summer, even in countries like the British Isles, watering can be an enormous help. Of course, it is essential in regions where summers are hot and dry. In a mixed border watering can be done with a hose pipe. In a rose garden or a rose border, or where other kinds of plants also require water, there is much to be said for some form of irrigation. If you do water, do it thoroughly, so that the moisture goes deep into the soil and reaches all the roots.

Growing English Climbers

Here are a few points that apply especially to English Climbing Roses. Their growing and maintenance is very much the same as for other English Roses, or indeed for other climbing roses.

PLANTING When planting against a wall, it is important to place your roses a little away from it, at about 45cm/18in distance, with their roots pointing away from the wall. The soil against the wall frequently remains dry and is impoverished, and it is important to enable the roots to reach fertile, moist soil as quickly as possible.

TRAINING In the first year simply tie back the branches to the wall, fence, trellis, pillar, arch or other support. In subsequent years, spread the growth out to cover the area as completely as possible. Having said this, you do not want a too-controlled appearance. There may be times, as the years go by, when certain branches create a wayward but pleasing effect and may be allowed to have their own way, to some extent.

The method used to tie in English Climbers depends on the structure to be clothed. On walls, growth can be kept in place by parallel wires set at intervals of about 46cm/18in and attached to vine eyes. Simply thread new growth behind and in front of the wires. Treat roses on trellis in the same way, occasionally threading the growth under the woodwork. With other forms of support, stems will require tying with twine or string. Ties should be loose enough to allow stems to expand.

MULCHING, FEEDING AND WATERING Careful attention to maintenance is particularly important where English Climbers are concerned, as a climbing rose has to produce a great deal of growth and bloom. Watering, in particular, should be extra-generous.

PRUNING Many climbers, of whatever kind, have a tendency to remain bushy for some time before beginning to climb. This is particularly true of English Roses, some varieties of which can be used either as a climber or as a shrub. We find it best to remove some of the short, twiggy branches at the base of the plant in order to concentrate all its strength into the long, climbing shoots. This is important only in the first year or two.

Once the climber is established, it is necessary to preserve the long, climbing shoots and cut back the short side branches, that have borne the flowers, to about three leaf buds. These will then produce next year's crop of flowers.

As the years go by and the main stems age or become too plentiful, some of the less productive growth should be removed altogether, leaving the vigorous, young growth.

Pests and Diseases

Roses in general certainly have their fair share of diseases. Insect pests are rather less of a problem, the main one being

aphids. These are easily controlled by spraying with a proprietary insecticide.

We have laid great emphasis on the breeding of disease-resistant English Roses, and have had a fair degree of success. If they are planted either singly, or in small groups, at intervals in mixed borders around the garden, rather than all together in one place, this will usually prevent cross-infection and it may well be unnecessary to spray them at all. Nonetheless, it is still advisable to keep a look out for disease and to take the necessary steps if it does occur— this is most likely to happen in a garden that is very enclosed. The best and easiest precaution you can take against disease is to plant roses that are naturally disease-resistant. Most English Roses introduced since 1983 are in this category and those introduced since 2000 are very resistant to disease.

Where English Roses are planted in close proximity—as in a rose border or rose garden—some spraying will be necessary. This is particularly true if they are mixed with roses of other groups that may well be less resistant to disease.

A precaution worth taking before the growing season begins is to rake up and remove all old and dying leaves and twigs that have fallen to the ground during the previous autumn and winter and while pruning. It is these that will carry disease forward into the coming year. If you then prick over the soil with a fork, this will not only help the growth of the roses but will also bury any spores left on the ground from the previous year. You will, of course, have removed any diseased shoots on the plant at pruning time.

There are four main diseases of roses.

MILDEW Most people will recognise this disease: it takes the form of white, powdery spores on the leaves. It is easily controlled. Ample watering will do much to avoid it.

BLACKSPOT As the name suggests, this appears in the form of large, black spots with uneven edges which, if

untreated, will eventually destroy the leaves and weaken the plant. It usually starts in midsummer (July onwards in Britain). Blackspot is encouraged by the leaves remaining wet for six hours or more when the weather is warm. If you water, do so early in the day so that the leaves can dry before the temperature rises.

RUST Rose rust produces rust-like protuberances, which gradually turn to black, on the underside of the leaf. It occurs as the weather becomes warmer in summer (late June and July in Britain).

DOWNY MILDEW This occurs when the nights are cold and the days are warm, usually either at the beginning of the flowering season or towards the end. It is not a common disease and you may never actually see it, but when it does occur, it is quite difficult to eradicate. It is also difficult to detect. Generally, the first sign is when the leaves start to fall for no apparent reason. However, if you look at the underside of the leaves, you will notice a very light, downy fungus. This may be so faint that you need a magnifying glass to see it.

SPRAYING Prevention is better than cure, so start spraying early in the season and you may find that you have to do very little for the remainder of the year. Choose a spray that cures all diseases or one that deals with the specific disease affecting your roses.

If you have a large number of roses, it is as well to use a knapsack sprayer; for just a few, a hand sprayer will be adequate. Always be careful to cover the whole surface of the leaves (including the underside) as well as the stems.

INDEX

Page references in **bold**
refer to illustrations

THROUGHOUT THIS BOOK I have referred to each variety by its commercial name. David Austin Roses Ltd reserves all Intellectual Property Rights on their rose varieties and trade marks are listed below. Throughout this book each variety is referred to by its commercial name (eg Heritage). The variety denomination (eg AUSBLUSH) of all varieties protected by Plant Breeders' Rights worldwide is clearly stated in Part Two on each variety's main descriptive page, but has been omitted elsewhere for clarity.

Trade marks

This book will be read in many countries, so below is a list of David Austin trade marks that have Trade Mark rights somewhere in the world. For a definitive list for a specific country, contact the Licensing Department, David Austin Roses Limited, on +44 1902 376327.

A Shropshire Lad
Abraham Darby
Admired Miranda
Ambridge Rose
Anne Boleyn
Austins Buttercup
Austins Herbalist
Austins Windflower
Barbara Austin
Belle Story
Benjamin Britten
Bibi Maizoon
Blythe Spirit
Bredon
Brother Cadfael
Buttercup
Charity
Charles Austin
Charles Darwin
Charles Rennie
 Mackintosh
Charlotte
Charmian
Christopher
 Marlowe
Claire Rose
Comtes de
 Champagne
 (syn. Coniston)
Coniston
 (syn. Comtes de
 Champagne)
Constance Spry
Cordelia
Corvedale
Cottage Rose
Country Living
Crocus Rose

Crown Princess
 Margareta
Eglantyne
Ellen
Emanual
Emily
English Elegance
English Garden
Evelyn
Fair Bianca
Falstaff
Fisherman's Friend
Francine Austin
Geoff Hamilton
Gertrude Jekyll
Glamis Castle
Golden Celebration
Grace
Graham Thomas
Happy Child
Harlow Carr
Heather Austin
Heavenly Rosalind
Heritage
Hyde Hall
James Galway
Janet
Jayne Austin
John Clare
Jubilee Celebration
Jude The Obscure
Kathryn Morley
L D Braithwaite
Leander
Lilic Rose
Lilian Austin
Lochinvar
Lucetta

Ludlow Castle
Malvern Hills
Marinette
Mary Magadalen
Mary Rose
Mary Webb
Mayor of
 Casterbridge
Miss Alice
Mistress Quickly
Molineux
Mortimer Sackler
Mrs Doreen Pike
Noble Antony
Othello
Pat Austin
Peach Blossom
Pegasus
Perdita
Portmeirion
Prospero
Queen Nefertiti
Radio Times
Redouté
Rose-Marie
Rosemoor
Rushing Stream
Scarborough Fair
Scepter'd Isle
Sharifa Asma
Shropshire Lass
Sir Edward Elgar
Snow Goose
Spirit of Freedom
St. Alban
St. Cecilia
St. Swithun
Sweet Juliet

Symphony
Teasing Georgia
Tess of the
 d'Urbervilles
The Alexandra Rose
The Alnwick® Rose
The Countryman
The Dark Lady
The Generous
 Gardener
The Ingenious
 Mr Fairchild
The Mayflower
The Pilgrim
The Prince
The Squire
The Yeoman
Tradescant
Trevor Griffiths
Troilus
Warwick Castle
Wenlock
Wife of Bath
Wildeve
William Morris
William Shakespeare
William Shakespeare
 2000
Winchester
 Cathedral
Windflower
Windrush
Wisley
Yellow Charles
 Austin
Austin
David Austin
David Austin Roses

Picture credits   *t* = top, *a* = above, *b* = below, *l* = left, *r* = right

CLAIRE AUSTIN: 60 *t*. BRIDGEMAN (www.bridgeman.co.uk): 4 *l* and 9 (Phillips, The International Fine Art Auctioneers, UK); 8 (Musée Archeologique, Sfax, Tunisia); 10 *a* (Österreichische Nationalbibliothek, Vienna); 10 *b* (Large Clive Album, Victoria & Albert Museum, London, The Stapleton Collection); 11 (Private Collection). FERUCCIO CARASSALE: 291. RON DAKER: 21; 62; 245–7; 257–8; 262 *a*; 264–5; 275; 293 *no.6*; 294 *a*; 207. ANDREW LAWSON: 1 ('Charles Rennie Mackintosh'); 2 ('The Countryman'); 23 *t*; 29; 31 *b*; 33; 34 *l*; 35 *all*; 39 *tr*; 243; 251; 253; 263; 266–7; 274; 292 *nos.2–5*; 293 *nos 8–9*; 295. ROGER PHILIPS: 23 *l*; 25 *a*. HOWARD RICE: 4 *r*; 5 *all*; 13–17 *all*; 20 *all*; 22 *all*; 23 *a*; 24 *all*; 25 *l*; 26–27 *all*; 30; 31 *t*; 32 *all*; 34 *r*; 36–38 *all*; 39 *t* and *l*; 40–59 *all*; 60 *a*; 61; 63–4; 71–201 *all*; 205; 209–237 *all*; 244; 248–50 *all*; 252; 254–6 *all*; 261; 262 *b*; 269–72 *all*; 276–89 *all*; 292 *no. 1*; 293 *no.7*; 294 *b*. UNKNOWN PHOTOGRAPHER: 12; POLLY WICKHAM/Land Art Ltd: 273

A FIREFLY BOOK

Published by Firefly Books Ltd. 2007

Text copyright © David Austin Roses Ltd 2007
Design and layout copyright © David Austin Roses Ltd 2007

Second printing, 2008

Publisher Cataloging-in-Publication Data (U.S.)

Austin, David, 1926–
    David Austin's English roses / David Austin
[304] p. : col. photos. ; cm.
Includes index.
Summary: Explores the new classifications developed by David Austin: The Old Rose Hybrids, small to medium-sized shrubs, with the charm of Old Roses enhanced by disease resistance, more continuous flowering and more bushy growth; The Aloha Group, tall, arching growth, mostly with large, deeply cut flowers; English-Alba Hybrids, tall, airy, natural growth, with almost Wild Rose daintiness; and The English Musk Roses, vigorous, bushy growth with reliable repeat flowers of particular delicacy.
ISBN-13: 978-1-55407-351-1 (bound)
ISBN-10: 1-55407-351-0 (bound)
ISBN-13: 978-1-55407-445-7 (pbk.)
ISBN-10: 1-55407-445-2 (pbk.)
1. English roses. 2. Roses-Varieties. I. II. Title.
635.933734 dc22   SB411.65.E53A978   2008

Library and Archives Canada Cataloguing in Publication

Austin, David, 1926–
    David Austin's English roses / David Austin.
Includes index.
ISBN-13: 978-1-55407-351-1 (bound)
ISBN-10: 1-55407-351-0 (bound)
ISBN-13: 978-1-55407-445-7 (pbk.)
ISBN-10: 1-55407-445-2 (pbk.)
    1. Old roses—Varieties. 2. English roses—Varieties.
I. Title. I. Title: English roses.
SB411.65.E53A88 2008   635.9'33734   C2007-903700-3

Published in the United States by
Firefly Books (U.S.) Inc.
P.O. Box 1338, Ellicott Station
Buffalo, New York 14205

Published in Canada by
Firefly Books Ltd.
66 Leek Crescent
Richmond Hill, Ontario L4B 1H1

Designed and typeset in Stempel Garamond by Ken Wilson
Jacket photography: Howard Rice
Back jacket arrangement: Shane Connolly

Printed in China